2-50

Fluoride, Teeth and Health

SUMMARY OF A REPORT ON
FLUORIDE AND ITS EFFECT ON
TEETH AND HEALTH FROM
THE ROYAL COLLEGE OF PHYSICIANS OF LONDON

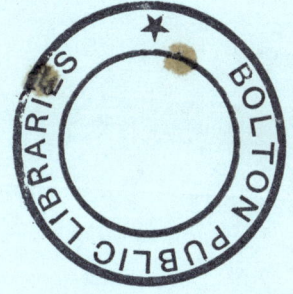

PITMAN MEDICAL

Summary

1. Introduction

Caries or dental decay is a disease involving cavity formation and irreversible destruction of teeth. It is basically of microbial origin but its prevalence is closely linked with diet and, particularly, with the consumption of refined carbohydrate.

It has been shown in many parts of the world that the amount of dental caries in the population varies inversely with the amount of fluoride in drinking water: that is, the higher the level in water, the less caries there is in the community. Fluoridation, the addition of fluoride to bring the concentration up to one milligram per litre, usually regarded as the optimal level, has therefore been widely recommended.

Only about half-a-million people in the United Kingdom drink water containing fluoride present naturally at a level of about 1 mg/litre, so that the opportunity exists of improving dental health on a large scale by fluoridation of water supplies. In fact, fluoridated water is available to less than 10 per cent of the population, due largely to the concern that has been expressed that the addition of fluoride might have some other effects on health which counterbalance the benefit to the teeth. It seemed, therefore, that it might be helpful if the Royal College of Physicians, as an independent body concerned with the public health, were to review the evidence.

2. Caries: The Size of the Problem

Caries is one of the commonest diseases in the world and indeed it may be the commonest of all chronic diseases in Europe. Because caries is so common and unspectacular it tends to be regarded as an inevitable part of life. Nevertheless it is hardly trivial.

By the age of 5 years, on average, every child has had four decayed deciduous (milk) teeth and by 15 years, on average, 10 of the 28 permanent teeth of every child are either decayed, missing or filled. Over 7,000 dentures are supplied to school children each year under the National Health Service.

Caries is the principal reason for dental extraction, and the large proportion of people in the community who have none of their own teeth is a striking indication of its impact. In 1968, 37 per cent of a sample of people over the age of 16 in England and Wales were found to have had all their teeth extracted, and in Scotland this proportion was still higher (44 per cent).

Dental disease is one of the most costly diseases to the nation, its treatment amounting to over £140,000,000 in England alone each year. It has been estimated that it is responsible for the loss of approximately 2 million working days each year.

It is easy to overlook the fact that caries has a small but appreciable mortality. Thus, it is responsible for some deaths from bacterial endocarditis, a condition to which people with valvular disease of the heart are liable when septic teeth are extracted. Each year, also, about 12 people die in England and Wales as a result of dental anaesthesia for which caries is a principal indication.

3. Fluoride and Caries

The existence of a connection between fluoride and dental decay was realised after it was established that high levels of fluoride in water caused mottling of the teeth. This discovery was made in the U.S.A. by McKay and others in 1931 and later confirmed in the high-fluoride town of Maldon, Essex, where it was also noted that the children there had less caries than the national average. Repeated studies confirming this finding led to the first study of fluoridation—the addition of fluoride to water to bring the concentration up to 1 mg per litre with the object of reducing the incidence of caries in the community. Surveys in Grand Rapids, U.S.A. and also in the nearby low fluoride community of Muskegon were carried out before fluoridation (to obtain baseline information) and again six-and-a-half years later. These showed that the amount of caries in 6-year-old children had been greatly reduced in Grand Rapids almost to half that of 6-year-old children in Muskegon. Later comparison showed that after 15 years of fluoridation the proportion of decayed, missing or filled teeth of 15-year-old children had fallen from 12·5 to 6·2 teeth.

Within a few years of the first results of these Michigan studies confirmation came from fluoridation studies in other parts of the U.S.A. and from Canada. Later, other countries introduced schemes, including the U.K., where fluoridation was introduced in Anglesey, Watford and Kilmarnock. The consistency of these many studies in indicating a marked reduction in the prevalence of caries is striking.

Several studies have shown that the benefits of fluoride are lifelong and are not restricted to childhood. Thus, less caries was found up to the age of 65 years in Hartlepool, Co. Durham, which has always had a high fluoride level in its water as compared with York, which has a low fluoride level (*see* Chapter 3 and Tables 3.1 to 3.3).

It has been repeatedly established that both children and adults in communities that consume water containing 1 mg per litre over the years of tooth formation (up to the age of 14 years) have a substantially lower prevalence of dental caries.

4. Objections to Fluoridation: A Synopsis

1. It has been suggested at different times that fluoride causes a wide variety of disorders, and these are listed on page 18, which is in the section of this report where the evidence is discussed. Apart from these specific disorders it is sometimes said that the fluoride added to water is different in some respect from that present naturally. However, the fluoride salts used in fluoridation dissolve, releasing fluoride ions similar in every respect to those already present. There is no difference in absorption of fluoride from soft water as opposed to hard water. (*See* Physiology, p. 23).

Another criticism is that there is not a hundred-fold margin between 1 mg per litre and the lowest levels at which toxic effects have been recorded. This margin, though essentially arbitrary, has been adopted as a guideline for considering additives to food but would be quite inappropriate in many other circumstances. Examples include water itself, and vitamin D.

It has also been argued that since the volume of water consumed by individuals in a community cannot be controlled, the consequent latitude in fluoride intake offends medical practice. However, the purpose of fluoridation is not to administer a specific dose to each person in the population but to replicate the effects observed in communities which already receive water with this concentration of fluoride. (*See* Chapter 3).

2. The suggestion that fluoridation is unnecessary since there are other equally or more effective methods of achieving the same result is discussed on page 11 in the Summary, together with the relative merits of other methods of giving fluoride.

3. The objection that fluoridation is uneconomic and wasteful is discussed on page 12 in the Summary.

4. The criticism that fluoridation is of negligible benefit is sometimes made, and is usually based on selecting data that refer to the age group 8 to 11 years in which caries mainly affects the surfaces least protected by fluoride: namely, the first molar and pre-molar fissures (*see* Chapter 3).

The statement is sometimes made that fluoridation merely postpones and does not prevent caries. Several factors contribute to the development of caries and the removal of one factor (a relative lack of fluoride) cannot be expected to prevent the others from having an effect in the course of time. This is the case in many diseases. The elimination of one factor among several may, nevertheless, be well worthwhile. It is certainly so in the case of fluoride intake and dental caries, since the adequate provision of fluoride reduces the prevalence of the disease throughout life.

5. A frequently heard objection is that even if fluoridation is beneficial and safe it encroaches on individual liberty. However those who put this objection to us accepted the regular addition of several other substances to drinking water such as copper sulphate, chlorine, aluminium and calcium. It is doubtful if the distinction between such substances and fluoride is a reasonable basis for regarding fluoridation as unwarranted. The Committee is however concerned with the propriety of withholding a procedure if this is safe and of benefit. As emphasised in Chapter 2, caries is not a trivial dis-

order but one that is responsible for a great deal of morbidity and for an appreciable number of deaths from dental anaesthesia and from bacterial endocarditis.

5. Physiology and Toxicology

Fluorine is an element that virtually never occurs naturally in its free, gaseous form. In the form of fluorides, however, it is one of the most plentiful and widespread of elements. Fluorides are to be found in water, soil, rocks and the air, besides most foods, many plants and in many animal tissues. Most water supplies contain small amounts in solution but in certain supplies the concentration has been as high as 5·8 mg/litre in Britain, 16 mg/litre in the U.S.A. and even 95 mg/litre in Africa.

The daily intake of fluoride from solid food is small in Britain (about 0·5 to 1·0 mg) but tea constitutes an important source. Tea contains an average of about 1–2 mg/litre if made from water containing a low level of fluoride, or 2–3 mg/litre if water is used containing 1 mg/litre; this means that one cup may contain, in these two situations, 0·33 and 0·50 mg respectively. Clearly, the total daily intake will depend upon the amount of water and tea that is drunk over the course of the day. A person drinking 20 cups of tea or one and a half gallons of water would receive less than 8 mg in a low fluoride area and less than 13 mg in an area where the level in water was 1 mg/litre.

Fluoride is mostly readily absorbed. Excretion is mainly in the urine, with about 10 per cent in faeces, and smaller amounts in sweat. It is taken up by the bones and teeth in which it is always to be found, even in low fluoride areas. Indeed the bones contain more than 90 per cent of the total fluoride in the body.

Although the question cannot be regarded as proven, there is increasing evidence that fluorine is essential for life and health, and recent work on animals indicates that it is necessary for fertility and growth. The World Health Organisation Expert Committee on Trace Elements in Human Nutrition has included fluorine in their list of trace elements believed to be essential for animal life.

Acute fluoride poisoning is a serious disorder and has occurred mainly as a result of the accidental contamination of food by fluoride preparations, though it has also been a method of suicide. The lethal dose of sodium fluoride is about 4 to 5 grams. It would clearly be impossible to ingest such an amount from water containing 1 mg of fluoride per litre since it would be necessary to drink, over a short period, 1,000 litres (over 200 gallons) of such water to receive even 1 gram of fluoride.

Some patients, who are given 9 to 54 mg of fluoride ion per day in the treatment of osteoporosis or Paget's disease, experience nausea, abdominal discomfort and vomiting, just after taking the tablets.

Long-continued exposure to excessive amounts of fluoride results in mottling of the teeth and skeletal changes. This was first recorded in men who inhaled fluoride dust during the extraction of aluminium from cryolite, a mineral containing fluoride. A similar disorder, named 'endemic fluorosis', is due to the consumption of water containing high levels of fluoride. This is often asymptomatic, and nearly all cases with symptoms have been in tropical parts of the world (*see* Chapter 7).

6. Dental Mottling

Mottling of enamel was the first effect of fluoride to be noted in man and it has received a great deal of attention. It should be stressed that there are many other causes of mottling and that the degree of mottling caused by fluoride has a wide range. In certain studies minor degrees, often detectable only using a strong light and a lens, have been termed 'mild fluorosis' which in some circumstances can be misleading since this term would not be applied to the same appearances in a low fluoride area.

There is a close relationship between the fluoride level in water and mottled enamel; the higher the fluoride level the greater the proportion of affected children. It has been shown in both the U.S.A. and the U.K. that at a level of 1 mg per litre there is no noticeable mottling. Indeed, at least two studies found *fewer* cases of even mild mottling at this level than in low fluoride areas.

7. Skeletal Effects

A. *Skeletal Fluorosis*

As mentioned in 5 (Toxicology), long-continued exposure to excessive quantities of fluoride in certain industrial processes or in drinking water may cause skeletal fluorosis. This is characterised by increased density of bones (osteosclerosis) and the formation of bony outgrowths (exostoses). Skeletal fluorosis is often unassociated with symptoms, but in severe cases pain and deformities may occur, sometimes with pressure effects on adjoining nerves or the spinal cord. These cases have mainly been recorded in India and other tropical areas and no case has been recorded from the U.S.A. or the United Kingdom from areas with a fluoride level in water of 1 mg per litre. Indeed, no symptomatic case has been recorded from these countries even from areas with a level of 8 mg/litre. Cases recorded in India at levels below 3 mg/litre probably reflect other factors such as a greater water intake and malnutrition. A further factor may be the consumption of sediment high in fluoride content which is known to be a feature of well water in areas of endemic fluorosis during periods of water shortage. Cases are mainly found where there has been continuous exposure to 20 to 80 mg of fluoride daily for ten to twenty years.

B. *Other Skeletal Effects*

Three studies in the U.S.A. found *less* osteoporosis (and, in one case, fewer fractures) in certain high fluoride areas than in areas where the level was lower. However, in these studies the high fluoride level was appreciably in excess of 1 mg/litre and a low incidence of osteoporosis has not been reported from areas with 1 mg/litre, with the exception of Watford shortly after it was fluoridated. In areas with high levels of naturally present fluoride, a lower mortality from falls and accidents has been found in the U.S.A., England and Wales, an effect that has been attributed to a lower fracture rate due to stronger bones in high fluoride areas.

C. Special Cases

Patients with chronic renal failure who are dialysed with fluid containing 1 mg of fluoride per litre receive an additional load of fluoride by an unusual route. It was reported from a dialysis centre in Canada that such patients were particularly liable to certain bone complications (renal osteodystrophy). Further work on this question has been conflicting. Thus, certain dialysis centres that receive fluoridated water encounter a few cases with this complication while other non-fluoridated centres have many cases. At least two studies investigating the problem suggested that fluoride was not implicated, while recently another study has produced evidence of a positive effect of fluoride on the progression of osteomalacia in patients on long-term haemodialysis. The increasing use of de-ionisers in such cases therefore seems prudent.

A few cases have been reported from the U.S.A. and Argentina in which X-rays signs of skeletal fluorosis developed in patients with chronic renal disease who drank excessive quantities of water containing 1·7 to 5·7 mg/litre fluoride. Only one of these patients had symptoms of fluorosis, a patient in Texas who drank water with a level of 2·2 to 3·5 mg/litre of fluoride for 43 years.

8. Renal Disorders

As fluoride is mainly excreted by the kidneys, the question has been raised as to whether these organs are at risk if the excretory load of fluoride is increased. However, no differences have been detected between high and low fluoride areas in the incidence or mortality of renal disease in Britain and the U.S.A. Renal disorders have not been recorded in patients with fluorosis in the Punjab, where the most intensive work on the disorder has been carried out.

9. Congenital Malformations

The suggestion that fluoride is a cause of mongolism was prompted by a study in which more cases of this condition were found in certain U.S. towns with high levels of fluoride in their water than in other towns with lower levels. The cases of mongolism were found by screening institutions together with birth and death certificates, but no account was taken of affected children living at home, apart from that small proportion of cases recorded on birth certificates. Not surprisingly, therefore, even the highest rates were lower than those generally reported from studies that involve an intensive case-finding effort. Careful case-finding studies, both in U.S.A. and Britain, have found no difference in the incidence of mongolism between high and low fluoride areas.

There is no evidence that the incidence of mongolism or any other congenital malformation is increased by fluoride.

10. Cancer

Cancer has been mentioned in connection with fluoride largely because of certain experimental work on cell cultures and on animals with transplanted cancer, some of which have reported stimulation and others inhibition of growth. Other experiments have not confirmed these effects.

Studies in the U.S.A. and Britain have found no relationship between cancer mortality or cancer incidence and fluoride levels in water supplies.

11. Other Conditions

A wide variety of disorders has been regarded by one author as evidence of allergy or intolerance to fluoride in water at a level of 1 mg/litre. Evidence of these effects has not been detected by other workers either in parts of Europe or the U.S.A. with levels of up to 8 mg/litre or in India with levels of over 11 mg/litre. Some patients with osteoporosis or Paget's disease treated with 20 to 120 mg of sodium fluoride daily experience gastrointestinal symptoms, just after taking tablets. However, there is a considerable difference in concentration between taking even a single sodium fluoride tablet (9 mg of fluoride ion) with one cup of water and the same amount of fluoride taken in about 2 gallons of water (containing 1 mg/litre) over many hours.

A connection between fluoride and the thyroid has been suggested but no increase has been found in the incidence of any thyroid disorder in areas with 1 mg of fluoride per litre in water relative to lower levels.

There is no evidence that any other disorder can be caused by fluoride in water at a level of 1 mg/litre or even at higher levels.

12. Fluoridation in Perspective

I. Present State of Fluoridation

Fluoridation schemes now function in more than 30 countries, serving a population of over 150 millions. In England and Wales approximately 4¾ million people receive fluoridated water and an additional half a million receive water containing a similar level of naturally present fluoride. Fluoridation has been the subject of official enquiries by government commissions and legal tribunals in Canada, New Zealand, Australia, South Africa and the Republic of Ireland. In all cases, after taking a great deal of evidence, the safety and efficacy of the procedure has been endorsed. In July 1969 the World Health Organisation, in its twelfth plenary meeting, adopted a resolution recommending Member States to fluoridate water supplies that contain low levels of fluoride. The resolution reads—

'To recommend Member States to examine the possibility of introducing and where practicable to introduce fluoridation of those community water supplies where the fluoride intake from water and other sources for the given population is below optimal levels, as a proven public health measure; and where fluoridation of community water supplies is not practicable to study other methods of using fluorides for the protection of dental health.'

This resolution has recently (1975) been reaffirmed by the World Health Organisation at its 13th Plenary meeting which observed that fluoridation 'remains the most effective known means of preventing dental caries' and noted 'the further scientific proof of the safe use of fluoride in its various forms.'

There has been much determined opposition to fluoridation for reasons that have already been considered and, in fact, in 1971 Sweden repealed the legislation of 1962 enabling local authorities to initiate fluoridation although a similar motion had been heavily defeated in 1968. There is evidence that the voting in 1971 was affected by party affiliations. In Denmark the prohibition of the addition of fluoride to food and water was made preparatory to legislation enabling the fluoridation of water supplies with a low fluoride content, but this legislation has not yet been introduced.

II. Fluoridation Compared to Other Methods of Reducing Caries

It is sometimes said that fluoridation is unnecessary since there are equally effective methods of achieving the same result. It is true that caries would decline if people stopped eating refined carbohydrates, but it cannot be expected that this would occur on a large scale. A number of alternative methods have been used for giving fluoride, both systemic and topical. Of the former the most well known is fluoride tablets. However, great motivation on the part of parents is required to ensure that children receive the proper dose over a period of years. The fluoridation of milk has been suggested and though its supply might be fairly straightforward to arrange at school, an important period of tooth formation takes place before school age.

Topical methods such as the application of gels and solutions by dental personnel are time-consuming and require individual actions while fluoride-containing toothpastes have only a limited effect. As community measures, tablets and topical methods are much less effective than fluoridation.

III. Environmental Aspects

The fluoride level in rivers would be still appreciably less than 1 mg/litre if all water supplies low in fluoride content were fluoridated. Natural levels of over 6 mg/litre are recorded in other countries. The level in the sea (0·8 to 1·4 mg/litre) would be little affected both because of the enormous dilution that would occur and because most of the fluoride used would be derived from sources that would otherwise have been discharged to the sea as waste. Fluoridation does not harm the environment.

IV. Economic Considerations

The annual cost of fluoridation per head of the population including an allowance for capital outlay for equipment is very small and in Birmingham a recent estimate was 1·5p. Several studies have attempted to measure the benefits, though certain of these, such as a reduction in the pain of caries, are impossible to quantify. Appreciable reductions have been recorded in the costs of dental treatment and in the staff required for the provision of a dental service which amply justify the cost of fluoridation. This saving is in spite of the fact that only a small proportion of a fluoridated water supply is consumed by children under the age of 14.

13. Conclusions

1. Fluoride in water added or naturally present at a level of approximately 1 mg/litre over the years of tooth formation substantially reduces dental caries throughout life.
2. The consumption of water containing approximately 1 mg/litre of fluoride in a temperate climate is safe irrespective of the hardness of the water.
3. In comparison with fluoridation, systemic fluoride supplements such as tablets, drops and fluoridised salt have not been shown to be as effective on a community basis.
4. Fluoridation does not harm the environment.

14. Recommendation of the College

The College recommends fluoridation of water supplies in the United Kingdom where the fluoride level is appreciably below 1 mg per litre.

Fluoride, Teeth and Health

Fluoride, Teeth and Health

A REPORT AND SUMMARY ON
FLUORIDE AND ITS EFFECT ON
TEETH AND HEALTH FROM
THE ROYAL COLLEGE OF PHYSICIANS OF LONDON

PITMAN MEDICAL

First published 1976

Pitman Medical Publishing Co Ltd
42 Camden Road, Tunbridge Wells,
Kent TN1 2QD

Associated Companies

UNITED KINGDOM
Pitman Publishing Ltd, London
Focal Press Ltd, London

USA
Pitman Publishing Corporation, New York
Fearon Publishers Inc, California

AUSTRALIA
Pitman Publishing Pty Ltd, Melbourne

CANADA
Pitman Publishing, Toronto
Copp Clark Publishing, Toronto

EAST AFRICA
Sir Isaac Pitman and Sons Ltd, Nairobi

SOUTH AFRICA
Pitman Publishing Co SA (Pty) Ltd, Johannesburg

ISBN: 0 272 79373 6
Cat. No. 21 3242 81

Text set in 11/12 pt. Photon Baskerville, printed by photolithography,
and bound in Great Britain at The Pitman Press, Bath

Preface

In October 1973 the Royal College of Physicians agreed to set up a special Committee in response to a request received from the dental profession for views on the merits of fluoridation of water supplies. The dental profession was apparently convinced of the value of fluoride in preventing dental caries, and considered that the opinion of a body such as the College on other more strictly medical aspects of fluoridation would be very helpful.

The composition of the College Committee on the Fluoridation of Water Supplies is as follows—

Sir Cyril Clarke (President and Chairman)
Sir Melville Arnott
Sir Richard Doll
Dr J. Vallance-Owen
Dr T. P. Mann
Dr W. J. E. Jessop
Dr A. S. V. Burgen
Dr W. H. Taylor
Dr R. Goulding
Dr L. J. Kinlen (Hon. Secretary)
Dr M. Bush*
Dr C. J. Roberts*
Dr P. M. Sheppard
Dr J. J. Murray
Dr M. N. Naylor

Sir Kenneth Robson (Registrar—retired 1975)
Dr D. A. Pyke (Registrar—appointed 1975)
Dr P. A. J. Ball (Assistant Registrar—retired 1975)
Dr J. S. Malpas (Assistant Registrar—appointed 1975)
Mr G. M. G. Tibbs (Secretary of the College)
Miss K. M. Snow (Committee Secretary)
Miss R. V. Flower (Committee Secretary)

* Representing the Faculty of Community Medicine.

Contents

1   Introduction

It has been shown in many parts of the world that the amount of dental caries in the population varies inversely with the amount of fluoride in drinking water. Fluoridation, the addition of fluoride* to bring the concentration up to one milligram per litre,† usually regarded as the optimal level, has therefore been widely recommended.

Only about half a million people in the United Kingdom drink water containing fluoride that is present naturally at a level of approximately 1 mg/litre, hence there is an opportunity to improve dental health on a large scale by fluoridation of water supplies. In fact, fluoridated water is available to less than 10 per cent of the population, due largely to the concern expressed that the addition of fluoride might have some other effects on health which would counterbalance the benefit to the teeth. It seemed, therefore, that it might be helpful if the College, as an independent body concerned with the public health, were to review the evidence.

Dental caries is an unspectacular disorder that tends to be regarded as an inevitable part of life simply because it is so common; nevertheless it is hardly trivial. At worst it is responsible each year for a number of deaths from anaesthetics given in the course of dental treatment and from bacterial endocarditis, to which people with valvular diseases of the heart are liable when septic teeth are extracted. Dental caries is also responsible for much pain and for the loss of over one million working days a year. Within the National Health Service the cost of dental treatment in England alone

---

* Fluoride here and throughout the Report refers to fluoride ions.
† Equivalent to one part per million.

is more than £140 million a year, most of which is spent on treating caries and its complications.

The Committee appointed by the College included representatives of the dental profession and specialists in general medicine, paediatrics, community medicine, toxicology, epidemiology and genetics. Over the past eighteen months this Committee has examined a mass of data relating to the effects of fluoride, and of fluoridation of water supplies in particular. Attention has been given not only to published work but also to certain work as yet unpublished to which the Committee has had access. Oral evidence was taken from leading members of organisations opposed to fluoridation and the literature published by these bodies was examined in detail. The Committee is most grateful to all those who helped its work and especially to its own Honorary Secretary Dr L. J. Kinlen.

2   Caries: The Size of the Problem

Dental caries is one of the commonest diseases in the world and, indeed, it may be the commonest of all chronic diseases in Europe. Caries is characterised by cavitation and irreversible destruction of deciduous and permanent teeth. Though basically of microbial origin its prevalance is closely correlated with dietary factors, particularly the use of refined carbohydrates. It is virtually unknown in animals in the wild though it can be induced under specific conditions in the rat, hamster, monkey and other mammals.

Because caries is so common it tends to be regarded as an inevitable part of life. By the age of 15, 97 per cent of children in Britain have had some caries and this proportion is 99 per cent in Norway where, by the age of 21, only one person in a thousand is free of the condition.[10] A survey in 1973 of 13,000 children in England and Wales[8] found that approximately 75 per cent of children aged 5 years had some caries or its sequelae and on average every child had four affected deciduous teeth. In spite of the considerable dental treatment these children had previously received, 20 per cent of those aged 5 to 8 years and 10 per cent of those aged 9 to 15 had five or more decayed teeth requiring treatment.[8] By the age of 15 years the average child has ten of the twenty-eight permanent teeth (excluding third molars) either decayed, missing or filled[5] while in the age group 16 to 34 this number is 16·4 in males and 17·8 in females. In 1972, 7·8 million courses of dental treatment were provided under the National Health Service for children of school age.[2]

Caries is one of the principal reasons for dental extractions and the large proportion of people in the community who have none of their own teeth is a striking indication of its impact. In 1968, 37 per cent of a carefully drawn random sample of people over 16 years in England and Wales were found to have had all their teeth extracted[4] and in Scotland this proportion was even higher, at 44 per cent.[9] It is, of course, impossible to measure the pain caused by caries but there can be no doubt that it is considerable. The contribution of dental problems to malnourishment in the elderly is also unknown but is probably appreciable.

The impact of dental caries is also reflected in the cost of its treatment and its effect on sickness absenteeism. Dental disease is one of the most costly diseases to the nation,[6] its treatment amounting to over £140 million a year in England alone.[1] It has been estimated that 0·6 million days work are lost each year for which claims are made for sickness incapacity benefit.[3] This does not include absenteeism for periods of less than three days, which must represent a considerable proportion of the total workdays lost on account of caries—probably in the region of two million days annually.

Because of the immense morbidity, it is easy to overlook the corresponding mortality. Caries certainly contributes to mortality from bacterial endocarditis, but other factors are involved and it is not possible to measure its contribution precisely. Each year, also, some people die as a result of dental anaesthesia and in England and Wales this number is about twelve each year.[7]

REFERENCES

1. Department of Health and Social Security (1975). *Annual Report for 1974*. London: H.M.S.O.
2. Department of Health and Social Security (1974). *Health and Personnel Social Services Statistics*. London: H.M.S.O.
3. Department of Health and Social Security (1975). Unpublished data (by permission).
4. Gray, P. G., Todd, J. E., Slack, G. L. and Bulman, J. S. (1970). *Adult Dental Health in England and Wales in 1968*. London: H.M.S.O.
5. *Health of the School Child 1966–1970* (1972). London: H.M.S.O.
6. Office of Health Economics (1969). *The Dental Service*. Publication No. 29. London: H.M.S.O.
7. Office of Population Censuses and Surveys (1975). Unpublished data (by permission).

8. Todd, J. E. (1975). *Childrens Dental Health in England and Wales in 1973.* London: H.M.S.O.

9. Todd, J. E. and Whitworth, A. (1974). *Adult Dental Health in Scotland in 1972.* London: H.M.S.O.

10. *World Health Organisation Chronicle* (1969), **23**, 505.

3   Fluoride and Caries

The discovery that fluoride reduced the incidence of caries arose from the finding that fluoride at certain concentrations caused mottling of the teeth.[39] Mottled enamel was described in 1901 in Italian children by Eager,[20] though the effect was not attributed to fluoride at the time. In 1916, McKay[29,30] working in Colorado and Texas, U.S.A. established that this mottling was initiated during the period of calcification of the teeth by an unidentified factor present in drinking water. Evidence that the causative agent was present in water came not only from noting that affected communities sometimes shared a common water source but also from the observation that the prevalence of mottling sometimes altered following changes in the water supply.[30–32,34,26] It was not until 1931 that Smith[41] and his colleagues in Arizona, Churchill[11] in Pennsylvania, and Velu[43] in Morocco independently demonstrated that the water in affected areas contained concentrations of fluoride as high as 13·7 mg/litre. This association was confirmed when it was shown that those communities in which there had been a change in the prevalence of dental mottling had in fact changed the fluoride content of their water supplies.[15]

In the United Kingdom, in 1933, Ainsworth[3] showed that dental mottling in Maldon in Essex was also associated with a high level of fluoride in its water (4·5 to 5·5 mg/litre) and noted, more significantly, that children in the town had a lower prevalence of caries in their permanent teeth (7·9 per cent) than a sample of children from England and Wales as a whole (13·1 per cent). This observation, which had also been implied by impressions recorded by McKay and Black,[33] was confirmed in the U.S.A. by Dean and his colleagues in a series of studies, involving thousands of children in several States.[13,15–17] This marked inverse relationship between the fluoride content of drinking water and experience of caries in the population is shown in Figure 3.1,[28] which also shows that

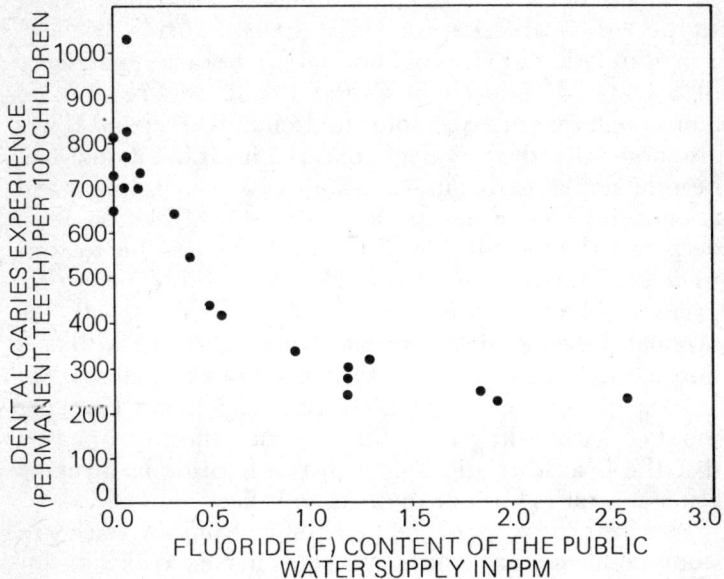

Figure 3.1   *Relation between the caries experience of 7,257 12- to 14-year-old white school children and the fluoride content of the water supply (Dean et al., 1942). Courtesy the Editor, Fluoride Drinking Waters (McLure, 1962).*[28]

above a concentration of 1 mg/litre very little further reduction is obtained. Sporadic instances only of the mildest form of dental mottling have been reported at this level, which required careful examination for their detection.[12,14]

PLANNED INTERVENTION
These studies formed the foundation of a crucial development: namely, the raising of the concentration of fluoride to approximately 1 mg/litre in water supplies with

low fluoride levels with the object of reducing the incidence of caries in the community. This procedure, that is, fluoridation, was first carried out in Grand Rapids, Michigan. In January 1945, after a dental survey of almost 20,000 children in this town, and over 4,000 children in the nearby low-fluoride town of Muskegon, to determine the pre-existing prevalence of caries, sodium fluoride was added to the water supply of Grand Rapids, raising its fluoride level to 1 mg/litre. After 6½ years of fluoridation the caries experience of 6-year-old children in Grand Rapids was reduced to almost half that of 6-year-old children in Muskegon.[4] It was also noted that the prevalence of caries in Grand Rapids had become similar to that in Aurora, Illinois, which already had a fluoride level of 1·4 mg/litre. Because Muskegon itself became fluoridated in July 1951, it was not possible to continue using this town as a control. However, subsequent comparisons of caries prevalence in Grand Rapids with the original baseline data indicated that after 15 years of fluoridation, the D.M.F. (decayed, missing or filled) score of 15-year-old children had fallen from 12·5 to 6·2 teeth per mouth.[6] Taken altogether, this evidence effectively proved that the inverse relationship between fluoride in drinking water and caries was one of cause and effect.

Within three years of the first results from the Michigan study, confirmation came from a study in New York State involving Newburgh, which was fluoridated in May 1945. Comparisons of caries experience in 1955 in children aged 10 to 15 with children of the same age in 1945 showed that the proportion of teeth decayed, missing or filled had fallen from 23·5 to 13·9 per cent. In contrast, in the non-fluoridated town of Kingston there was a slight increase in this proportion from 23·1 to 26·3 per cent.[7] Similarly, in Evanston, Illinois, which was fluoridated in January 1946, the mean number of decayed, missing or filled teeth in 14-year-olds fell from 11·7 to 6·0 between 1946 and 1960, a reduction of 49 per cent, whereas no change was observed in 14-year-old children in the nearby control town of Oak Park.[8]

In Britain, Ainsworth's[3] earlier finding of a reduced caries experience in the natural high-fluoride town of Maldon in

Essex (over 4 mg of fluoride per litre) had been confirmed by Weaver's observations in South Shields, Co. Durham.[44] During the Second World War, Irvine, a school dentist in Westmorland, noted that children evacuated to his area from South Shields had remarkably good teeth—much better than those of the local children. He mentioned this impression to Weaver who determined that the fluoride level in the water in this town was about 1·4 mg/litre. Weaver took the matter further and found in a study involving 2,000 children that the mean D.M.F score at the age of 5 years was 3·9 compared to 6·6 in North Shields on the other side of the Tyne where the fluoride level was much lower (0·25 mg/litre). Moreover, at the age of 12, the number of D.M.F. teeth in South Shields was 56 per cent of that in North Shields.

In 1955, following a mission from the British Government to the U.S.A. and Canada to study the practice in operation, fluoridation was carried out in Britain in Watford, Kilmarnock and part of Anglesey. Surveys five years later[18] showed that caries experience in deciduous teeth in the fluoridated areas had dropped markedly compared with the initial baseline levels, whereas in the control areas there had been little change. These beneficial effects were further confirmed in a survey eleven years after the introduction of fluoridation.[19] Even in Kilmarnock, where the Borough Council had discontinued the scheme in 1962 because of local opposition, a reduced prevalence of caries was still observed in certain age groups. Kilmarnock, incidentally, provided an opportunity to observe the changes that occur in caries experience after stopping fluoridation. It was found that children born after the scheme was stopped had an experience of caries similar to that of children of corresponding age in the control (low-fluoride) town of Ayr. It may be noted that a survey[36] had indicated that in fact a majority of the Kilmarnock residents did not support the discontinuation. More recent studies in Britain, in Birmingham (which was fluoridated in 1964) and Anglesey,[24] have also found a reduced prevalence of caries.

Several studies have shown that the dental benefits of fluoride are lifelong and not restricted to childhood.[1,2,10,21,22,35,40] Murray[38] found that in Hartlepool, Co.

Durham, which has always had a high fluoride level in its water (1·5 to 2·0 mg/litre), the mean D.M.F. score was consistently lower at all ages up to 65 than in York which has a low fluoride level (0·15 to 0·25 mg/litre) (*see* Figure 3·2). The prevalence of decayed or filled surfaces in adults aged 45 or over in Hartlepool was little more than half (55 per cent) that in York.

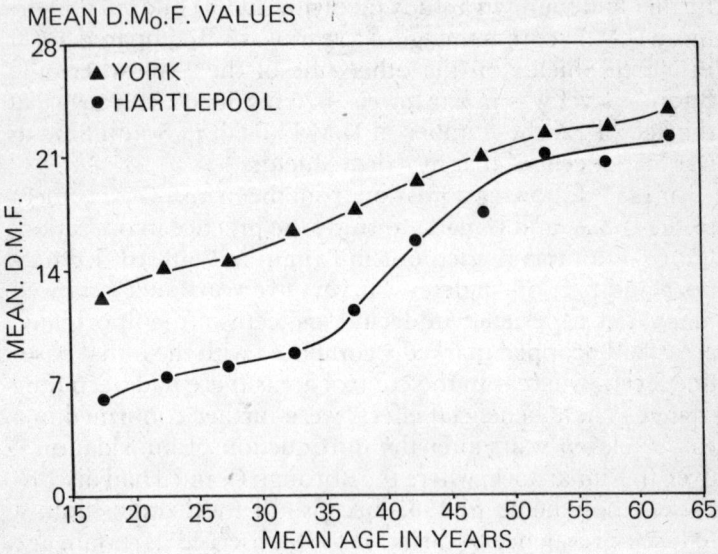

MEAN D.M$_O$.F. VALUES

▲ YORK
● HARTLEPOOL

Figure 3.2   *Mean number of decayed, missing, and filled permanent teeth in dentate adults from Hartlepool (1·5–2·0 p.p.m. F) and York (0·15–0·25 p.p.m. F). Courtesy* the Editor, *British Dental Journal.*

Data concerning the efficacy of fluoridation are usually expressed in terms of the number of decayed, missing or filled (D.M.F.) teeth or tooth surfaces. Other measures may be used, including the proportion of persons who either are free of caries or have experienced extensive disease. Tables 3.1 and 3.2 show the results of certain U.S.[5,7] and U.K.[19] studies of fluoridation, and of naturally present fluoride,[37] in terms of the proportion of people who are entirely free of caries, a stringent test of efficacy since even a single recognisable cavity is sufficient to remove an individual from the caries-free category. Table 3.3, A, B and C show the effect of fluoride expressed in terms of the proportion of people who have ten or more decayed teeth.[19,37] Whichever measure is employed, the beneficial effect of fluoridation in reducing the incidence of caries is apparent. Fluoridation studies in Canada,[9,23] the Netherlands,[27] Finland, Ireland, New Zealand, the U.S.S.R., Puerto Rico and South America[46] have also testified to the beneficial effects of the procedure on dental health.

TABLE 3.1

U.S.A. Proportion of children caries-free in fluoridated and control communities (after 10 years fluoridation)

A. *Grand Rapids—Muskegon Study (Permanent teeth only)*

Age—Years	Grand Rapids (Fluoridated) Caries free	Muskegon (Non-fluoridated) Caries free
6	89·3%	79·8%
7	66·8%	49·7%
8	49·4%	27·5%
9	33·1%	14·5%
10	26·6%	5·7%
Total 6–10	2,871	1,423

(Arnold *et al.*, 1956)[5]

B. *Kingston—Newburgh Study (Deciduous canines and molars)*

Age—Years	Newburgh (Fluoridated) Caries free	Kingston (Non-fluoridated) Caries free
6	37·0%	11·1%
7	27·9%	4·7%
8	24·9%	1·8%
9	10·1%	1·6%
Total 6–9	734	940

(Ast *et al.*, 1956)[7]

TABLE 3.2

U.K. Proportion of children caries-free before and after fluoridation

A. *Watford–Sutton Study.*[19] Children aged 3–7 years. Deciduous teeth

	Watford (fluoridated in 1956)	Sutton (non-fluoridated)
1956	19% (631)	25% (449)
1965	50% (677)	38% (665)

B. *Watford–Sutton Study.*[19] Children aged 8–10 years. Permanent teeth

	Watford (fluoridated in 1956)	Sutton (non-fluoridated)
1956	14% (560)	14% (474)
1967	42% (345)	23% (254)

C. *Anglesey Study.*[19] Children aged 3–7 years. Deciduous teeth

	Gwalchmai (fluoridated in 1956)	Holyhead (fluoridated in 1956)	Bodafon (non-fluoridated)
1955–56	13% (1,397)	13% (1,125)	12% (1,289)
1965	32% (744)	40% (554)	14% (628)

D. *Hartlepool–York Study.*[37]

Age	Hartlepool (high-fluoride) Caries-free	York (low-fluoride) Caries-free
5 years	51·2% (500)	22·4% (527)
15 years	6·2% (386)	0·8% (381)

Numbers in parentheses refer to the numbers of children examined.

TABLE 3.3.

Proportion of children with 10 or more decayed teeth

A. *Watford–Sutton Study.*[19] *(Ages 3–7 years. Deciduous teeth only)*

	Watford (fluoridated in 1956)	Sutton (non-fluoridated)
1956	11% (449)	8% (631)
1964	1% (665)	3% (677)

B. *Anglesey Study.*[19] *(Ages 3–7 years. Deciduous teeth only)*

	Gwalchmai (fluoridated in 1956)	Holyhead (fluoridated in 1956)	Bodafon (nonfluoridated)
1956	18% (1,397)	16% (1,125)	16% (1,289)
1965	5% (774)	2% (554)	13% (628)

C. *Hartlepool—York Study.*[37]

Age	Hartlepool (High-fluoride)	York (Low-fluoride)
5 years	1·6% (500)	9·9% (527)
15 years	8·7% (386)	39·9% (381)

Numbers in parentheses refer to the numbers of children examined.

Fluoridation studies throughout the world are strikingly consistent in showing a reduced prevalence of caries. However, the effect is not equal at different ages and at ages 8 to 11 years, it is relatively small. This is because at these ages caries occurs mainly in the occlusal pits and fissures of newly-erupted pre-molars and molar teeth which are less protected by fluoride than are smooth surfaces. Thus, if the data from this age group is singled out from the 1969 Department of

Health study,[19] it appears that fluoridation merely postpones caries by 0·8 cavities a year. It is relevant that at these ages less than half the permanent teeth have erupted and of these, caries chiefly attacks only the first four molar teeth; consequently a reduction of one in the D.M.F. score would reflect a substantial change. The good effect can be obscured further by grouping 11-year-old children, who have always consumed fluoridated water, with older children whose teeth were already partly formed when fluoridation was introduced. Such limited comparisons have usually formed the basis of the criticism sometimes made that fluoridation is of negligible benefit.

MODE OF ACTION

The precise mechanism by which fluoride reduces caries is not understood. It has been suggested that the effect is due to one or more of the following properties of fluoride:[25]

1. The reduced solubility of enamel formed during exposure to fluoride, perhaps due to the conversion of some hydroxyapatite into less soluble fluorapatite.
2. The remineralisation of enamel that fluoride promotes.
3. An antibacterial action of fluoride for which there is suggestive evidence.

The concentration of fluoride in enamel is higher in the outer layers where the caries process is less rapidly progressive than in the deeper layers, a microcosm of the epidemiological relationship.

Dental decay can be reduced by limiting the intake of refined carbohydrate or total food consumption. However, in both these situations there is evidence that fluoride exerts an additional protective influence.[42,44,45]

It has been repeatedly established that both children and adults in communities that consume water containing 1 mg of fluoride per litre over the years of tooth formation (up to the age of 14 years) have a substantially lower prevalence of dental caries.

REFERENCES

1. Adler, P. (1951). *Dtsch. Zahn-, Mund- u. Kieferheilk.*, 15, 24–30. Quoted in W.H.O. 1970.
2. Adler, P. (1953). *Schweiz. Mschr. Zahnheilk*, 63, 432–452. Quoted in W.H.O. 1970.
3. Ainsworth, N. J. (1933). *Brit. dent. J.*, 55, 233.
4. Arnold, F. A. Jun., Dean, H. T. and Knutson, J. W. (1953). *Publ. Hlth. Rep. (Wash.)*, 68, 141.
5. Arnold, F. A. Jun., Dean, H. T., Jay, P. and Knutson, J. W. (1956). *Publ. Hlth. Rep. (Wash.)*, 71, 652.
6. Arnold, F. A., Jr., Likins, R. C., Russell, A. L. and Scott, D. B. (1962). *J. Amer. dent. Ass.*, 65, 780.
7. Ast, D. B., Smith, D. J., Wachs, B. and Cantwell, K. T. (1956). *J. Amer. dent. Ass.*, 52, 314.
8. Blayney, J. R. and Hill, I. N. (1967). *J. Amer. dent. Ass.*, 74, 225.
9. Brown, H. K. and Poplove, M. (1965). *Med. Services J. of Canada*, 21, 450.
10. Bruszt, P. (1962). *Forgorv. Szle*, 55, 102–111. Quoted in W. H. O. 1970.
11. Churchill, H. V. (1931). *Ind. Eng. Chem.*, 23, 996.
12. Dean, H. T. (1936). *J. Amer. med. Ass.*, 107, 1269.
13. Dean, H. T. (1938). *Publ. Hlth. Rep. (Wash.)*, 53, 1443.
14. Dean, H. T. and Elvove, E. (1936). *Amer. J. pub. Hlth.*, 26, 567.
15. Dean, H. T. and McKay, F. S. (1939). *Amer. J. pub. Hlth.*, 29, 590.
16. Dean, H. T., Jay, P., Arnold, F. A. Jr. and Elvove, E. (1939). *Publ. Hlth. Rep. (Wash.)*, 54, 862.
17. Dean, H. T., Arnold, F. A. Jr. and Elvove, E. (1942). *Publ. Hlth. Rep. (Wash.)*, 57, 1155.
18. Dept. of Health (1962). *Rep. on Publ. Hlth. & Medical Subjects No. 105*. London: H.M.S.O.
19. Dept. of Health (1969). *Rep. on Publ. Hlth. & Medical Subjects No. 122*. London: H.M.S.O.
20. Eager, J. M. (1901). *Publ. Hlth. Rep. (Wash.)*, 16, 2576.
21. Englander, H. R. and Wallace, D. A. (1962). *Publ. Hlth. Rep. (Wash.)*, 77, 887.
22. Forrest, J. R., Parfitt, G. J. and Bransby, E. R. (1951). *Mon. Bull. Min. Hlth. Pub. Hlth. Lab. Serv.*, 10, 104.
23. Hutton, W. L., Linscott, B. W. and Williams, D. B. (1951). *Canad. J. Pub. Hlth.*, 42, 81.
24. Jackson, D., James, P. M. C. and Wolfe, W. B. (1975). *Brit. dent. J.*, 138, 165.
25. Jenkins, G. N. (1975). *Brit. med. Bull.*, 31, 142.
26. Kempf, G. A. and McKay, F. S. (1930). *Pub. Hlth. Rep. (Wash.)*, 45, 2923.
27. Kwant, G. W., Houwink, B., Backer Dirks, O., Groeneveld, A. and Pot, T. (1973). *Neth. Dent. J. Suppl. (No. 9)*, 6.
28. McClure, F. J. (Ed.) (1962). *Fluoride Drinking Waters.* U.S. Dept. of Health, Education & Welfare, Nat. Inst. of Dental Research, Bethesda, Maryland, p. 12.

29. McKay, F. S. (1916a). *Dent. Cosmos*, **58,** 477.
30. McKay, F. S. (1916b). *Dent. Cosmos*, **58,** 781.
31. McKay, F. S. (1918). *J. nat. dent. Ass.*, **5,** 721.
32. McKay, F. S. (1925). *Dent. cosmos.*, **67,** 847.
33. McKay, F. S. (1928). *J. Amer. dent. Ass.*, **15,** 1429.
34. McKay, F. S. (1933). *J. Amer. dent. Ass.*, **20,** 1137.
35. McKay, F. S. (1948). *Amer. J. Pub. Hlth.*, **38,** 828.
36. Martin, F. M. (1975). Unpublished data. Personal communication (July).
37. Murray, J. J. (1969). *Brit. dent. J.*, **126,** 352; **127,** 128.
38. Murray, J. J. (1971). *Brit. dent. J.*, **131,** 391, 437, 487.
39. Murray, J. J. (1973). *Brit. dent. J.*, **134,** 247, 299, 347.
40. Russell, A. L. and Elvove, E. (1951). *Pub. Hlth. Rep. (Wash.)*, **66,** 1389.
41. Smith, M. C., Lantz, E. M. and Smith, H. V. (1931). *Arizona University Agricultural State, Technical Bulletin,* **32.**
42. Tank, G. and Storvick, C. A. (1965). *J. Am. dent. Assoc.*, **70,** 394.
43. Velu, H. (1931). *C.R. Soc. Biol.*, **108,** 750.
44. Weaver, R. (1944). *Brit. dent. J.*, **76,** 29.
45. Weaver, R. (1950). *Brit. dent. J.*, **88,** 231.
46. *World Health Organisation Chronicle* (1969). **23,** 505.

4   *Objections to Fluoridation: A Synopsis*

Objections to fluoridation have been made on the grounds that it is dangerous, unnecessary, uneconomic or of negligible benefit; or that, even if safe and beneficial, it is unethical.

1. THE OBJECTION THAT FLUORIDATION IS DANGEROUS

It has been suggested at different times that many different disorders can be caused or aggravated by fluoridation. These disorders are listed on the next page with a reference to the chapters in this Report in which the evidence relating to them is discussed. There is a wide variation in the basis of these claims, which sometimes overlooks the difference between inorganic and organic fluoride or between fluoride in water at a level of 1 mg/litre and the concentrations used in certain *in vitro* or animal experiments. Many of the disorders listed below have been reported by only two authors, Waldbott and Spira.

Apart from the objections concerning specific disorders, certain more general criticisms are sometimes raised. Thus, it has been said that fluoride added to water differs from that which is present naturally. However, the soluble fluoride salts used in fluoridation dissociate, releasing fluoride ions similar in every respect to those already present. It has also been said that the calcium present in hard water binds fluoride ions so that less fluoride is absorbed than from soft water containing equivalent amounts. In fact, there is no difference in absorption of fluoride from soft water as opposed to hard water (*see* Chapter 5, p. 23).

Another criticism of fluoridation is that there is not a hundred-fold margin between 1 mg/litre and the lowest levels at which toxic effects have been recorded. This 'safety

Disorders claimed to be caused or aggravated by fluoridation (*Numbers refer to Chapters*).

Gastrointestinal disorders:	flatulence 5, 11(A)	diarrhoea 11(A)
	nausea 5, 11(A)	constipation 11(A)
	abdominal pain 5, 11(A)	gingivitis 11(A)
	vomiting 5, 11(A)	stomatitis 11(A)
	haematemesis 11(A)	oral ulcers 11(A)
	peptic ulcer 11(D)	
Neurological and mental disorders:	headache 11(A)	mental deterioration 11(A)
	migraine 11(A)	convulsions 11(A)
	depression 11(A)	personality change 11(A)
	paraesthesiae 11(A)	deafness 11(D)
	painful numbness in the limbs 11(A)	
Urinary tract disorders:	urethritis 11(A)	nephritis 8
	cystitis 11(A)	
	pyelitis 11(A)	
Skin disorders:	urticaria 11(A)	furunculosis 11(C)
	rashes 11(C)	brittle nails 11(A)
	dermatoses 11(C)	alopecia 11(A)
Musculo-skeletal disorders:	skeletal fluorosis 5, 7(A)	arthritis 7(B)
	backache 7(B)	renal osteodystrophy 7(C)
	pain on muscular assertion of the ribs (sic) 11(A)	Scheuermann's disease 7(B)
Dental disorders:	mottling of the teeth 6	
Ocular disorders:	optic neuritis 11(D)	
Endocrine disorders:	thyroid enlargement 11(B)	parathyroid disorders 11(C)
	diabetes mellitus 11(C)	adrenal disorders 11(C)
Cardiovascular disorders:	heart disease 11(D)	
	arteriosclerosis 11(D)	
Congenital malformations:	mongolism 9	
	anencephalus 9	
	spina bifida 9	
Cancer	10	
Non-specific disorders:	lethargy 11(A)	
	exhaustion after sleep 11(A)	
	muscular weakness 11(A)	

margin', though essentially arbitrary, has been adopted as a guideline for considering food additives and contaminants of food. Such a margin is quite inappropriate in many other instances, particularly for essential components of the diet. Examples include water itself and vitamin D. The recommended daily intake of vitamin D is $10 \mu g$, and the lowest recorded dose producing symptoms is $12 \mu g$.[1]

It has also been pointed out that since the volume of water consumed by individuals in a community cannot be controlled, the consequent latitude in fluoride intake offends normal medical practice. However, the purpose of fluoridation is not to administer a specific dose to each person in the population, but, as described in Chapter 3, to replicate the beneficial effects observed in communities that receive water with naturally present fluoride at a concentration of 1 mg/litre.

The criticism that fluoridation would constitute pollution of the environment is discussed in Chapter 12 (section III).

2. THE OBJECTION THAT FLUORIDATION IS UNNECESSARY
The suggestion that fluoridation is unnecessary because there are other equally or more effective methods of achieving the same result is discussed in Chapter 12 (section II), together with the relative merits of other methods of administering fluoride.

3. THE OBJECTION THAT FLUORIDATION IS WASTEFUL AND UNECONOMIC
This is discussed in chapter 12 (section IV).

4. THE OBJECTION THAT FLUORIDATION IS OF NEGLIGIBLE BENEFIT
This objection is discussed in Chapter 3; it is usually based on limited data referring to the age group in which caries mainly affects those surfaces least protected by fluoride, namely the first molar and pre-molar fissures.

It is also sometimes said that fluoridation merely postpones and does not prevent caries. A similar criticism could be levelled against the advice to stop smoking on the grounds that the risks of lung cancer, coronary thrombosis and chronic bronchitis increase with age and reach levels in non-smokers similar to those in cigarette smokers who are 20 years younger. Many factors contribute to the development of many diseases and the removal of one factor cannot be expected to prevent the others from having an effect in the course of time. The elimination of one factor may nevertheless be well worthwhile. It is certainly so in the case of fluoride intake and dental caries, since the adequate provision of fluoride reduces the prevalence of the disease throughout life.

5. THE OBJECTION THAT FLUORIDATION IS AN UNWARRANTED COMPULSORY MEASURE

A frequently heard objection is that even if fluoridation is beneficial and safe it encroaches on individual liberty. However those who put this objection to us accepted the regular addition of several other substances to drinking water such as copper sulphate, chlorine, aluminium and calcium. It is doubtful if the distinction between such substances and fluoride is a reasonable basis for regarding fluoridation as unwarranted. The Committee is however concerned with the propriety of withholding a procedure if this is safe and of benefit. As emphasised in Chapter 2, caries is not a trivial disorder but one that is responsible for a great deal of morbidity and for an appreciable number of deaths from dental anaesthesia and from bacterial endocarditis.

REFERENCE

1. Seelig, M. S. (1969). *Ann. N.Y. Acad. Sci.*, 147, 537.

5 The Physiology and Toxicology of Fluorides

PHYSIOLOGY

Fluorine is an element that virtually never occurs naturally in its free, gaseous form. In the form of fluorides, however, it is one of the most plentiful and widespread of elements, standing seventeenth in order of abundance in the earth's crust.[67] Fluorides occur in water, soil, rocks, dusts, volcanic gases and the atmosphere. They are also present in most foods, many plants and virtually all animal tissues.[13] Most water supplies contain small amounts of fluoride in solution but the concentration may be as high as 5·8 mg/litre in Britain,[24] 16 mg/litre in the U.S.A.[67] and even 95 mg/litre in Africa;[67] in sea water the level is 0·8 to 1·4 mg/litre.[67] High-fluoride fresh waters are often hard but there are many exceptions. Fluorides are also present in the atmosphere of most urban areas as a result of the combustion of coal and other fuels, and sometimes as a result of particular industrial processes such as aluminium extraction or the manufacture of steel using fluorspar.

INTAKE

Almost every foodstuff contains at least a trace of fluoride but the total intake from solid food is relatively small, and amounts to only about 0·5 to 1·0 mg/day in Britain.[7,33] High concentrations in fish are often quoted but because the fluoride is mainly in the bones and skin its importance as a source of dietary fluoride has been exaggerated. Tea contains more fluoride than any other dietary item in Britain. Dry tea leaves usually contain up to about 200 mg/kg,[67] much of which is extracted during infusions. The amount present in one cup of tea will depend not only on the size of the cup, but the brand of tea, the amount used, the duration of the infusion and whether it is from a dilution of a previous brew. Most published estimates of the average fluoride content of tea infusions have been about 1 mg/litre[33,67] though 3 mg/litre has been reported.[16,17] The approximate range of daily fluoride intake by adults in areas with water-fluoride levels of 0·1 and 1·0 mg/litre respectively is shown in Table 5.1. The fluoride content of tea infusions has here been assumed to be 2 mg/litre in a low-fluoride area and 3 mg/litre in an area with 1 mg/litre in its water supply. It is difficult to estimate an upper limit for the intake of tea or other fluids. The highest intake of tea recorded in a survey in Newcastle-upon-Tyne (which is fluoridated) was by an elderly man who drank 22 cups daily providing, with other sources, a total fluoride intake of about 9 mg per day.[32,33] In this case, the fluoride content of tea was lower than the estimate used in Table 5.1. Some individuals may obtain similar amounts of fluoride from beer or other fluids. It has been assumed in the table that a 'maximum' intake of tea would be associated with a corresponding reduction in the amounts of other drinks. Fluoride may also be absorbed by inhalation, but the contribution of atmospheric fluoride to intake is small in

TABLE 5.1

Estimated Daily Fluoride Intake by Adults (mg) (approx.)

Fluoride Level in Water	0·1 mg/litre Intake		1·0 mg/litre Intake	
	'Low-normal'	'Maximum'	'Low-normal'	'Maximum'
Food	0·50	1·00	0·70	1·20
Water (exc. tea)	0·10[1]	0·60[2]	1·00[1]	6·00[2]
Tea	0·66[3]	6·60[4]	1·00[3]	10·00[4]
Total	1·26	7·70[5]	2·70	12·20[5]

Footnotes:
1. Assuming a daily intake of 1 litre of water (excl. tea).
2. Assuming a daily intake of 6 litres of water (excl. tea).
3. Assuming a daily intake of 2 cups of tea, equivalent to $\frac{1}{3}$ litre.
4. Assuming a maximum of 20 cups of tea.
5. Assuming that a maximum tea consumption is associated with a 'low-normal' intake of water apart from tea.

Britain and is less than 0·1 mg/day.[15] It can be deduced from Table 5.1 that where there is little fluoride in the water the average daily intake by adults is less than 3·0 mg with a maximum of less than 8·0 g, while with 1 mg/litre in water, the average daily intake is less than 5·0 mg with a maximum of about 12 mg.

ABSORPTION

Fluoride is absorbed by a passive process from the alimentary tract. Radioactive tracer experiments have shown that maximum plasma levels are reached within an hour of ingesting 1 mg of fluoride.[12] With few exceptions, such as the bones of tinned fish, most of the fluoride in food, water and tea is absorbed. Calcium and magnesium in relatively large amounts have been shown in experiments to bind fluoride, but the amounts present in even the hardest water in Britain do not appreciably reduce absorption of fluoride present in water at a level of 1 mg/litre.[8] The aluminium, which is often added to water in its purification process, forms complexes with fluoride but without impairing its absorption.[8,32]

EXCRETION

Over 80 per cent of the total fluoride excreted by the body is in urine with about 10 per cent in faeces and smaller amounts in sweat, except during excessive perspiration, when appreciable amounts may be lost by this route.[36] Only traces are present in saliva, tears and milk. Urinary excretion is rapid. If a small dose of fluoride is taken in a glass of water, 20 per cent can be found in the urine within three hours.[12,25] The proportion of absorbed fluoride excreted in the urine depends on the extent of retention by the bones and teeth. This, in turn, is affected by age and the previous intake of fluoride.[35,36] Clearance by the kidneys is lower than that of inulin, suggesting that some tubular reabsorption occurs. The relationship first reported by McClure and Kinser between the fluoride concentrations in drinking water and in urine is strikingly close.[40]

BONE

Fluoride is taken up by the bones and the teeth, replacing certain ions normally associated with hydroxyapatite crystallite.[26,64,65] Concentration in bone increases with age.[28,52] Jackson and Weidmann[28] reported that a plateau was reached around the age of 55 years, the level depending on fluoride intake (Figure 5.1). This was not confirmed by Weatherell's[63] work on the femur (Figure 5.2). In West Hartlepool, which has a water–fluoride level of 1·5 to 2·0 mg/litre, the plateau level in rib was about 4,000 mg/kg compared to 2,000 mg/kg in Leeds where the fluoride level in water is 0·1 mg/litre. The concentration is higher in the areas of active growth, near the endosteal and periosteal surfaces, than in the central parts of compact bone.

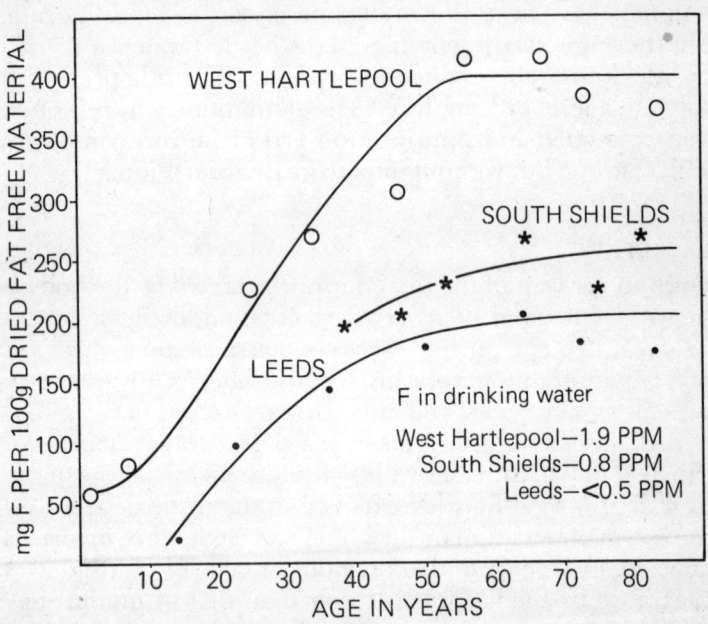

Figure 5.1   Fluoride content in post-mortem samples of human rib bone in areas with different levels of fluoride in the water supply. (Courtesy Jackson and Weidman (1958), J. Path. Bact., 76, 461.)

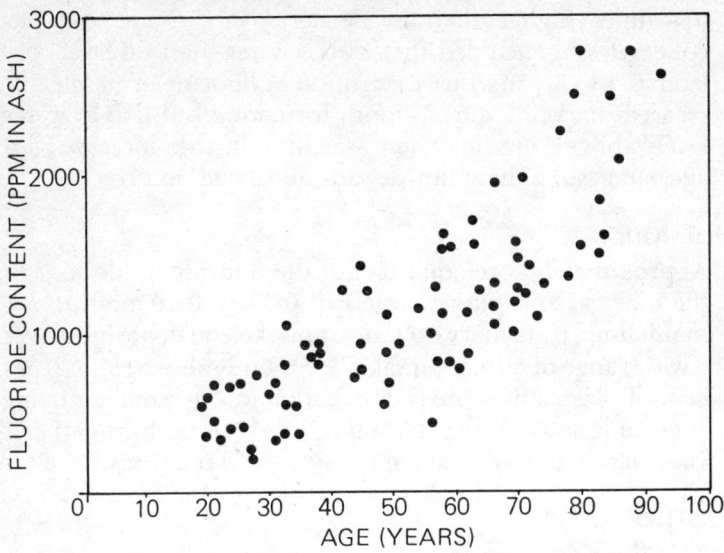

Figure 5.2   Fluoride content of femoral compacta from humans of different ages living in districts supplied with drinking water containing <0·5 p.p.m. F. (After Weatherell, 1966.)

TEETH

Fluoride concentration in teeth is lower than in bone and the increase that occurs with age is less marked. As with bone, it is influenced by the fluoride level in drinking water. Jackson and Weidman[29] found that in West Hartlepool, the content of premolar enamel rose from 170 mg/kg at 10 years to a plateau level of 300 mg/kg at 30 years. In Leeds, a 'low fluoride' city, the corresponding levels were 50 and 100 mg/kg. The fluoride concentration in premolar dentine, which in general is four times that of enamel, was found in West Hartlepool to increase from 400 mg/kg at 10 years to 1,200 mg/kg at 55 years, whereas in Leeds the level at 10 years was 100 mg/kg and 400 mg/kg at 55 years. Similar results have been reported by other workers.[3,30,31] After the tooth has fully formed fluoride is chiefly incorporated at tissue surfaces and its concentration in dentine is highest adjacent to the odontoblastic layer.[69] In the outer level of enamel it may be ten times higher than in deeper layers. Isaac and her colleagues[27] concluded that as the water fluoride level rose from 0·1 to 5·1 mg/litre deposition of fluoride in enamel increased markedly during tooth formation but that at water levels above 5 mg/litre there was little further increase with age, suggesting that fluoride saturation had occurred.

BLOOD

Approximately three quarters of the fluoride in blood is in the plasma, and plasma levels of 0·14 to 0·19 mg/litre are maintained by urinary excretion and skeletal deposition over a wide range of fluoride intake.[12,50,67] Only about 15 to 20 per cent of plasma fluoride is free and ionic, the remainer being non-ionic and bound to albumin.[11,51,56] It is the former that is considered to participate in physiological reactions.

OTHER TISSUES

The fluoride concentration in human milk and saliva is 0·05 mg/litre or less.[67] The concentration in other soft tissues is usually less than 1 mg/kg except in tissues such as the aorta[10,53] and the placenta in late pregnancy[53] which may contain high levels.

PREGNANT WOMEN AND CHILDREN

The placenta acts as a partial barrier to the passage of fluoride to the fetus, except when the fluoride level is low.[19,20,70] Urinary excretion is lower at this time and in the puerperium, returning to previous levels two or three months after delivery. Children excrete a smaller proportion of a given dose of fluoride[23] than adults, probably because they are actively laying down bone mineral for which fluoride has an affinity.

FLUORINE AS AN ESSENTIAL ELEMENT

The wide distribution of fluorine in animal tissues and particularly in bone and teeth has raised the question of whether it has a physiological role or whether its presence is accidental because it is ingested in food. However, its widespread presence in foodstuffs and water has made it difficult to investigate the question of whether it is essential for life and health. No human diet is free of fluoride, and even in the laboratory there are major problems in preparing, for animals, a diet that is very low in fluoride. The seven controlled studies of the question have had varying degrees of success.[18,38,39,42,43,44,48,49,62] In the most recent of these, Messer and his colleagues demonstrated that fluorine satisfied the major criteria for an essential trace element in the mouse. Female mice maintained on a low fluoride diet over two generations showed a progressive decline in litter production, whereas mice receiving the same diet supplemented with fluoride reproduced normally and at constant intervals. This impaired reproductive capacity was prevented and cured by the addition of fluoride to the diet. It is also of interest that Schwarz and Milne[48] showed that addition of 2·5 to 7·5 mg/kg of fluoride to a highly purified amino-acid diet containing 0·04 to 0·46 mg/kg fluoride significantly enhanced the growth rates of young mice.

In this connection it may be noted that the Food and Nutrition Board of the National Research Council of the U.S.A.[45] include fluoride in their recommended dietary constituents, and the U.S. Food and Drug Administration lists fluorine as an essential nutrient.[60] The WHO Expert Committee on Trace Elements in Human Nutrition[68] also included fluorine in their list of 14 trace elements believed to be essential for animal life.

TOXICOLOGY

ACUTE TOXICITY

Acute fluoride poisoning in man has mainly occurred as a result of the accidental contamination of food by sodium fluoride or sodium silicofluoride preparations; less often it has been a method of suicide. It has been estimated that in the 85 years prior to 1957, 435 cases of acute fluoride poisoning had been reported, of which over 300 occurred in two epidemics of food contamination.[67]

The most prominent changes in fatal cases are the presence of congestion and submucosal haemorrhages in the stomach and upper small intestine. It is noteworthy that animal experiments have shown that these changes are not due to local effects since they are as common after intravenous injections as after ingestion. While it is often difficult to estimate accurately either the amount taken or, because of vomiting, the amount retained, the data suggest that the median lethal dose of sodium fluoride is about 4 to 5 grams, approximately equivalent to 63 mg/kg.[21,47,61] The corresponding data for a considerable number of laboratory and farm animals have been reported and most fall within the range 42 to 210 mg/kg.[55,61] Gettler and Ellerbrook[21] estimated the fluoride concentration in soft tissues in fatal cases and found values of 19 to 28 mg/kg (excluding the brain) compared with normal values of 0·55 to 1·43 mg/kg, an approximately 25-fold increase.

It would clearly be impossible for water containing 1 mg/litre to cause the lethal effects described above, since it would be necessary to drink, over a short period, 1000 litres of such water to receive even 1 g of fluoride.

The symptoms are varied but the most prominent are gastro-intestinal, with abdominal pain, diarrhoea, salivation, nausea and vomiting followed by neurological effects such as painful spasms, paraesthesiae, weakness, paresis, convulsions and stupor. Dyspnoea is frequent and in some cases hypocalcaemia and tetany have been reported.[47,67]

A number of reports are available of the effects of smaller amounts ingested by human beings particularly since the advent of fluoride therapy in osteoporosis and Paget's disease. Bredemann[6] found that 250 mg (equivalent to 4·2 mg/kg) produced burning pains in the abdomen and vomiting. Griebel *et al.*[22] found that 100 mg (equivalent to 1·7 mg/kg) produced nausea and vomiting and 140 mg (equivalent to 2·5 mg/kg) vomiting, diarrhoea and severe abdominal pain. Black and Kleiner[5] found nausea and vomiting with doses of 30 to 80 mg sodium fluoride (0·4 to 1·3 mg/kg). Similarly, some patients with osteoporosis or Paget's disease treated with a daily dose of 20 to 120 mg of sodium fluoride equivalent to 9 to 54 mg of fluoride ion) experience nausea, epigastric discomfort and, in some cases, vomiting just after taking their tablets.[34,37,46,57,58] It is difficult to evaluate reports of aching joints or arthralgia in some of these patients[4,34,37] since such symptoms are common in these disorders irrespective of treatment. In this connection, the marked difference in concentration may be noted between taking even a single sodium fluoride tablet (9 mg of fluoride ion) in, say, one cup of water, and the same amount of fluoride in 9 litres of water (containing 1 mg/litre) taken over a few days or at least over many hours (*see* Chapter 11).

CHRONIC TOXICITY

In animals. Chronic toxic effects attributed to fluoride were first documented by Velu in North Africa in 1931 in animals feeding on herbage and drinking water contaminated with dusts from rock phosphate deposits and mine workings.[47] In Britain, fluorosis has been a problem in cattle herds in certain areas in which their pastures have been polluted with fluorides from neighbouring factories;[2,9] in Australia, the U.S.A. and Africa, sheep and cattle have become affected from drinking water high in fluoride content, while in Iceland, contamination of herbage by volcanic deposits has also resulted in fluorosis in these animals.[59] The signs of fluorosis in animals include loss of appetite, dental defects and lameness due to ankylosis and the formation of exostoses.[47,59]

In man. In man, long continued exposure to excessive amounts of fluoride is liable to result in skeletal changes and, if exposure occurs during the period of tooth development, in dental mottling. Skeletal fluorosis was first recorded in men who inhaled fluoride dust during the extraction of aluminium from cryolite (a mineral containing fluoride). In Britain, the only significant group of cases occurred in an aluminium factory near Fort William,[1] where many of the workers showed radiological signs of osteosclerosis but none had any physical disability or symptoms. The first descriptions of industrial fluorosis were soon followed by reports from India of a similar disorder named endemic fluorosis due to the consumption of water containing high levels of fluoride (*see* Chapter 7). In Spain, a related skeletal disorder has been attributed to an excessive intake of wine containing large amounts of sodium fluoride which had been added as a preservative.[54]

Although many conditions in man have been linked with fluoride in water, only dental mottling and skeletal fluorosis have been shown to be caused by fluoride. These are considered further in Chapters 6 and 7 respectively. There is no evidence, however, that either condition can be produced by fluoride at a concentration of 1 mg/litre of water, irrespective of whether the water is soft or hard. The evidence for other conditions being connected with fluoride is discussed in the subsequent chapters.

REFERENCES

1. Agate, J. N., Bell, G. H., Boddie, G. F., Bowler, R. G., Buckell, M., Cheeseman, E. A., Douglas, T. H. J., Druett, H. A., Garrad, J., Hunter, D., Perry, K. M. A., Richardson, J. D. and Weir, J. B. de V. (1949). United Kingdom Medical Research Council Memorandum No. 22, *Industrial Fluorosis*. London: H.M.S.O.

2. Allcroft, R., Burns, K. N. and Herbert, C. N. (1965). *Fluorosis in Cattle: 2. Development and Alleviation*. London: H.M.S.O. (Animal Diseases Survey, No. 2).

3. Armstrong, W. D. and Singer, L. (1963). *J. dent. Res.*, 42, 133.

4. Bernstein, D. S. and Cohen, P. (1967). *J. Clin. Endocrinol. and Metab.*, 27, 197.

5. Black, M. M. and Kleiner, I. S. (1947). *Cancer Research*, 7, 818.

6. Bredemann, G. (1956). *Biochemie und Physiologie des Fluors und der Industriellen Rauschaden*. Berlin: Akademie Verlag.

7. Burt, B. A. *et al.* (1973). *Brit. dent. Journal*, 135, 543.

8. Brudevold, F., Moreno, E. and Bakhos, Y. (1972). *Archs. Oral Biol.*, 17, 1155.

9. Burns, K. N. and Allcroft, R. (1964). *Fluorosis in Cattle: 1. Occurrence and Effects in Industrial Areas in England and Wales*. London: H.M.S.O. (Animal Diseases Survey, No. 2).

10. Call, R. A., Greenwood, D. A., Le Cheminant, W. H., Shupe, J. L., Nielsen, H. M., Olsen, L. E., Lamborn, R. E., Mangelcon, F. L. and Davis, R. V. (1965). *Publ. Hth. Rep. (Wash.)*, 80, 529.

11. Carlson, C. H., Armstrong, W. D. and Singer, L. (1960). *Amer. J. Physiol.*, 199, 187.

12. Carlson, C. H., Armstrong, W. D. and Singer, L. (1960). *Proc. Soc. exp. biol. Med.*, 104, 235.

13. Carlson, C. H., Singer, L. and Armstrong, W. D. (1960). *Proc. Soc. exp. biol. Med.*, 103, 418.

14. Caruso, F. S., Maynard, E. A. and Distefano, V. (1970). *Handbook of Experimental Pharmacology*, 20(2), 144.

15. Collings, G. H., Fleming, R. B. L. and May, R. (1951). *A.M.A. Arch. industry. Hyg.*, **4**, 585.

16. Cook, H. A. (1970). *Fluoride*, **3**, 12.

17. Cook, H. A. (1975). Personal communication, August.

18. Doberenz, A. R., Kurnick, A. A., Kurtz, E. B., Kemmerer, A. R. and Reid, B. L. (1964). *Proc. Soc. exp. Biol. (N.Y.)*, **117**, 689.

19. Feltman, R. and Kosel, G. (1955). *Science*, **122**, 560.

20. Gardner, D. E., Smith, F. A., Hodge, H. C., Overton, D. E. and Feltman, R. (1952). *Science*, **115**, 208.

21. Gettler, A. O. and Ellerbrook, L. (1939). *Amer. J. med. Sci.*, **197**, 625.

22. Griebel, C., Schleomer, A. and Zeglin, H. (1948). *Z. Untersuch. Lebensmittel*, **75**, 305.

23. Ham, M. P. and Smith, M. D. (1954). *J. Nutr.*, **53**, 215.

24. Heasman, M. A. and Martin, A. E. (1962). *Mon. Bull. Min. Hlth.*, **21**, 150.

25. Hodge, H. C. and Smith, F. A. (1965). Biological effects of inorganic fluorides. In: Simons, J. H. (ed.) *Fluorine Chemistry*. Vol. 4. New York: Academic Press, pp. 137.

26. Hodge, H. C. and Smith, F. A. (1968). *Ann. Rev. of Pharmacology*, **8**, 395.

27. Isaac, S., Brudevold, F., Gardner, D. E. and Smith, F. A. (1958). *J. dent. Res.*, **35**, 420.

28. Jackson, D. and Weidmann, S. M. (1958). *J. Path. Bact.*, **76**, 451.

29. Jackson, D. and Weidmann, S. M. (1959). *Brit. dent. J.*, **107**, 303.

30. Jenkins, G. N. and Spiers, R. L. (1954). *J. dent. Res.*, **33**, 734.

31. Jenkins, G. N. (1963). *J. dent. Res.*, **42**, 444.

32. Jenkins, G. N. and Edgar, W. M. (1973). *J. dent. Res.*, **52**, 984.

33. Jenkins, G. N. (1975). *British Med. Bulletin*, **31**, 142.

34. Jowsey, J., Riggs, B. L., Kelly, P. J. and Hoffman, D. L. (1972). *Amer. J. Med.*, **53**, 43.

35. Largent, E. J. (1959). In: Muhler, J. C. and Hine, M. K. (ed.) *Fluorine and Dental Health*, Bloomington, Indiana. Univ. Press, p. 132.

36. Largent, E. J. (1961). *Fluorosis: The health aspects of fluorine compounds*. Columbus: Ohio State Univ. Press.

37. Lukert, B. P., Bolinger, R. E. and Meek, J. C. (1967). *J. Clin. Endocrinol. and Metabolism*, **35**, 387.

38. McClendon, J. F. (1944). *Fed. Proc.*, **3**, 94.

39. McClendon, J. F. and Gershon-Cohen, J. (1953). *J. Agric. Food. Chem.*, **1**, 464.

40. McClure, F. J. and Kinser, C. A. (1944). *Pub. Hlth. Rep. (Wash.)*, **59**, 1575.

41. Machle, W. and Largent, E. J. (1943). *J. indus. Hyg.*, **25**, 112.

42. Maurer, R. L. and Day, H. G. (1957). *J. Nutr.*, **62**, 561.

43. Messer, H. H., Armstrong, W. P. and Singer, L. (1972). *Science*, **177**, 893.

44. Muhler, J. C. (1954). *J. Nutr.*, **54**, 481.

45. Jl. National Academy of Sciences (1974). *Recommended Dietary Allowances*, pp. 98 and 126. Food and Nutrition Board, Nat. Res. Council, Washington D.C.

46. Purves, M. J. (1962). *Lancet*, **ii**, 1188–1189.

47. Roholm, K. (1937). *Fluorine Intoxication*. London: Lewis.

48. Schwarz, K. and Milne, D. B. (1972). *Bio inorg. Chem.*, **1**, 331.
49. Sharpless, G. R. and McCollum, E. V. (1933). *J. Nutr.*, **6**, 163.
50. Singer, L. and Armstrong, W. D. (1960). *J. appl. Physiol.*, **15**, 508.
51. Singer, L. and Armstrong, W. D. (1959). *Anal. Chem.*, **31**, 105.
52. Smith, F. A., Gardner, D. E. and Hodge, H. C. (1953). *Fed. Proc.*, **12**, 368.
53. Smith, G. A., Gardner, D. E. and Hodge, H. C. (1960). *A.M.A. Arch. indust. Hlth.*, **21**, 330.
54. Soriano, M. (1965). *Revista Clinica Espanola*, **XCVII**, 375.
55. Spector, W. S. (1956). *Handbook of Toxicology*. Philadelphia: Saunders.
56. Taves, D. R. (1966). *Nature (Lond.)*, **211**, 192.
57. Taylor, W. H. (1970). *Brit. med. J.*, **4**, 304.
58. Taylor, W. H. (1975). Unpublished data. Personal communication.
59. Underwood, E. J. (1971). *Trace Elements in Human and Animal Nutrition (Fluorine)*. New York and London: Academic Press.
60. U.S. Federal Register (1973), 20713, No. 148, Aug. 2nd 1973. Superintendent of documents, Government Printing Office, Washington D.C. (FDA Regulation 125·1).
61. Waldbott, G. L. (1963). *Acta med. Scand.*, **174**, Suppl. 400, p. 1.
62. Wallace-Durbin, P. (1954). *J. dent. Res.*, **33**, 789.
63. Weatherell, J. A. (1966). *Handb. exp. Pharmak.*, **20**, 141.
64. Weidmann, S. M. and Weatherell, J. A. (1959). *J. Path. Bact.*, **78**, 243.
65. Weidmann, S. M. and Weatherell, J. A. (1970). *Fluorides and Human Health, 1970*. Geneva, p. 104.
66. Wiseman, A. (1970). *Handbook of Experimental Pharmacology*, **20**(2), 48.
67. World Health Organisation (1970). *Fluorides and Human Health*, Geneva.
68. World Health Organisation (1973). Technical Report No. 532. Geneva.
69. Yoon, S. H., Brudevold, F., Gardner, D. E. and Smith, F. A. (1960). *J. dent. Res.*, **39**, 845.
70. Ziegler, E. (1956). *Mitt. Naturw. Ges. Winterthur*, **28**, 1.

6 *Dental Mottling*

It was the finding that high levels of fluoride in water caused mottling of the teeth which led, as described in Chapter 3, to the discovery that fluoride reduces the incidence of caries. This mottling, or dental fluorosis, affects the permanent dentition and only rarely are deciduous teeth involved. In Britain, it was well described by Ainsworth in 1933[1] in children in Maldon, Essex, at a time when the fluoride level in water was 4·5 to 5·5 mg/litre. At such levels, lustreless white patches are liable to occur on the surface of the teeth, together with some yellow or brown staining. In severe cases, pitting may be present due to loss of enamel. However, it needs to be stressed that at lower levels mottling is usually minor and detectable only by experienced observers using a strong light and lens. Moreover, such mottling is not only caused by fluoride but also by many other factors that can affect tooth development. Indeed, a survey in low fluoride parts of Suffolk and Surrey showed that 63 per cent of children had enamel defects and that 6 per cent had marked brown staining.[3] Had these been noted in a high fluoride area they might easily have been termed 'mild fluorosis'. This fact underlines the importance of control groups in studies of mottling in relation to fluoride.

The relationship between the fluoride level in drinking water and the prevalence of mottled enamel was first investigated by Dean in the U.S.A. in the 1930s.[4,5] He found that the proportion of children with mottled enamel increased with increasing concentration of fluoride in their drinking water. Dean also showed that a fluoride level of 1 mg/litre was not associated with noticeable mottling although at levels of 2·9 mg/litre and over a high prevalence was noted. In 1944 this was confirmed in Britain by Weaver[10] who also found that the prevalence of mottling was similar in South Shields with a fluoride level of about 1·4 mg/litre, and North Shields where the level was approximately 0·25 mg/litre.

Dean noted, as have other observers since, that a fluoride level of 1 mg/litre was associated with teeth that had not only less caries but also a whiter and more uniform appearance. In fact at least two studies have found a lower prevalence of mottling at this level than in low fluoride areas. This was the finding of Forrest and James[6] in their study of 8-year-old children on Anglesey and it was confirmed more recently by Al-Alousi[2] and his colleagues in children aged 12 to 16 in Anglesey and Leeds, a low-fluoride city. At least one mottled incisor tooth was noted in 39 per cent of Anglesey children and 52 per cent of Leeds children, while 9 per cent of all incisor teeth in Anglesey were mottled as compared with 12 per cent in Leeds. No differences were found in another study between the prevalence or type of mottling in 5-year-old and 15-year-old children in Anglesey and the low-fluoride towns of Bangor and Caernarvon.[9]

In Arizona,[7] where the mean temperature is high and the fluid intake greater than in temperate climates, mottling has been reported in association with fluoride concentrations of over 0·8 mg/litre.

Two children with nephrogenic diabetes insipidus and dental mottling, aged 10 and 11 years respectively, have been reported[8] from the U.S.A. from areas with a fluoride level of 1 mg/litre. The daily fluid consumption of these children ranged from $2\frac{1}{2}$ to 6 times the normal daily intake and it was suggested that the fluoride present in the water may have caused the mottling. However, as the authors also observed, there were other causes for mottling in the past history of these children—fevers and convulsions. It is relevant that since the introduction of fluoridation in 1964, no cases of dental mottling have come to the notice of medical and dental staff in the Birmingham centre which specialises in the treatment of children with chronic renal failure, many of whom have polydipsia. Children who are considered for renal transplantation in this centre are routinely referred for a dental opinion.[11]

There is no evidence that in a temperate climate water containing fluoride at a concentrationof 1 mg/litre is associated with an increased prevalence of dental mottling.

REFERENCES

1. Ainsworth, N. J. (1933). *Brit. dent. J.*, **55**, 233.
2. Al-Alousi, W., Jackson, D., Crompton, G., Jenkins, O. C. (1975). *Brit. dent. J.*, **138**, 9.
3. Burt, B. A., Jackson, D., Jenkins, G. N., Murray, J. J. and Young, M. A. (1973). *Brit. dent. J.*, **135**, 543.
4. Dean, H. T. (1936). *J. Amer. med. Ass.*, **107**, 1269.
5. Dean, H. T. and Elvove, E. (1936). *Amer. J. Pub. Hlth.*, **26**, 567.
6. Forrest, J. R. and James, P. M. C. (1965). *Advanc. Fluorine Res.*, **3**, 319.
7. Galagan, D. J. and Lamson, G. G. (1953). *Public Hlth. Rep.*, **68**, 497.
8. Greenberg, L. W., Nelson, C. E. and Kramer, N. (1974). Pediatrics, **54**, 320.
9. Jackson, D., James, P. M. C. and Wolfe, W. B. (1975). *Brit. dent. J.*, **138**, 165.
10. Weaver, R. (1944). *Brit. dent. J.*, **76**, 29.
11. White, R. H. R. (1975). Personal communication (July).

7 Skeletal Effects

A. SKELETAL FLUOROSIS

Chronic exposure to excessive quantities of fluoride produces changes in the skeleton that were first described more than 40 years ago in workers who extracted aluminium from cryolite.[10,30,42] A similar disorder, but caused by fluoride in water and termed endemic fluorosis, was reported from the region of Madras in India in 1937[44,45] and later from other parts of the world; these have been mainly tropical areas. Skeletal fluorosis is characterised by radiological changes, particularly osteosclerosis, in association with a high fluoride concentration in bones. The osteosclerosis particularly affects the vertebrae and pelvis and is often associated with irregular osteophyte formation, increased thickness of long bones and calcification of ligaments and tendons. Histologically there is disturbance of the Haversian system, and formation of irregular osteoid seams and new bone, with some bone resorption in other areas.[9,15] Indeed, a very mixed picture may be presented with features suggestive of osteomalacia, Paget's disease, osteosclerosis and osteoporosis. Although fluorotic bone is sclerotic it is not as strong, weight for weight, as normal bone.

This type of skeletal fluorosis in Danish cryolite workers and in villagers in India is characterised by pain and limitation of movement of the vertebral column and the lower limbs, associated in some cases with deformities such as kyphosis, contracture of the hips or knees and fixation of the thoracic cage. In severe cases there may also be a radiculomyelopathy, a complication that has been recorded almost exclusively in India. In Texas, U.S.A., radiological evidence of fluorosis in the form of osteosclerosis has been recorded in 10 to 15 per cent of people studied in an area with a water–fluoride level of 8 mg per litre,[23] and X-ray changes have been noted in a few people in Oklahoma and Texas at levels of 4 to 8 mg/litre.[55] All these people however were symptom-free and had no physical disability. Similarly, the Scottish aluminium workers with fluorotic osteosclerosis who were investigated by a Medical Research Council group[2] were also free of symptoms. Other studies in the U.K.[8] and U.S.A.[27] in areas with up to 6 mg/litre of fluoride in water supplies have failed to detect even asymptomatic cases of skeletal fluorosis while in the U.S.S.R.[20] no radiological differences were detected between residents of areas with levels below 1 mg/litre and areas with 4 mg or more per litre.

In India, Jolly and his colleagues have observed from their considerable experience of over 1000 cases of skeletal fluorosis in the Punjab that severe cases result from the continuous exposure of 20 to 80 mg of fluoride daily for 10 to 20 years, associated with levels in water of at least 10 mg/litre.[15] In Madras, it was noted that such cases were seen only in persons who have continuously resided for at least 25 years in the high-fluoride areas, whereas in Hyderabad much shorter intervals were recorded for immigrants to an area where the fluoride level in wells was 9·2 to 11·8 mg/litre.[46] A few cases of skeletal fluorosis in children have been reported from an endemic area with fluoride levels in the range 10·4 to 13·5 mg/litre.[58] However cases with symptoms have also been reported in association with lower fluoride levels,[5,21,33,41,51–53,56] even in the range 1 to 3 mg per litre.[15,36,50]

In these tropical studies, no measurements have been made of fluoride intake other than that in water though several workers have commented on the large amount of sediment present in well water in hot weather conditions. Reference has also been made to the use of fluoride-containing stones for grinding food and condiments, besides brackish water of unknown fluoride content for cooking.[36] The existence of other sources is further suggested in certain cases by the presence of fluoride in urine in appreciably greater concentrations than in drinking water.[50] This may explain certain cases of fluorosis associated with relatively low levels of soluble fluoride. It is also relevant that cases of endemic fluorosis are usually related to the fluoride level in the villages where the patients currently live and this may not be appropriate. As the disorder mainly affects males it is likely that, when working, many men will have used wells other than those in their home villages. Moreover Shortt and his colleagues[44,45] in their first description of the disease noted that in the Madras region it was common for villagers to move from site to site.

Whatever the explanation, no symptomatic case of skeletal fluorosis attributable to water consumption has been recorded from areas with water-fluoride levels below 4 mg/litre except in tropical countries where, apart from a high fluid intake, the possibility of nutritional factors exists or of ingesting sediment with a high fluoride content. Only one case of non-industrial skeletal fluorosis has been recorded in Britain and this was associated with a myelopathy in a man aged 57 years.[61] The source of the excessive intake of fluoride in this case was not established though suspicion rested on the garden well (since blocked up) from which he had drunk water for five years from the age of 33. However, attempts to obtain a sample of this water to determine its fluoride content were unsuccessful. The authors also noted that the patient was a heavy tea-drinker.

The levels of fluoride in bone associated with the changes described above have not been clearly established.[54,63,64,65] The human skeleton has a high degree of physiological tolerance to the accumulation of fluoride and some workers have found that at concentrations below 2,500 mg/kg no abnormality can usually be detected in bone. Indeed, osteosclerosis was not apparent radiographically in certain studies below levels of 5,000 to 6,000 mg/kg of dry fat-free bone, or 10,000 mg/kg of bone ash.[54,62,64] The ribs of two workers with industrial fluorosis contained 8,500 and 10,000 mg/kg fluoride of bone ash.[42] Lower levels were reported by Singh and Jolly in ten cases of endemic fluorosis in India, ranging from 700 to 6,800 mg/kg compared with 200–300 mg/kg in people in a low fluoride area.[49]

B. OTHER SKELETAL EFFECTS

Apart from fluorosis, certain other skeletal conditions have been related to fluoride or the lack of it. In 1943, Leone and his colleagues[23] X-rayed the dorso-lumbar spine and pelvis of 116 residents, aged 15 to 78 years, in Bartlett, Texas (with a water fluoride level (until 1952) of 8 mg/litre, and 121 residents of the nearby town of Cameron (0·4 mg/litre of fluoride). Ten years later the X-rays were repeated on 89 of the original Bartlett group and 101 of the Cameron group. By 1953, 8 persons in the low-fluoride community had developed osteoporosis compared to 1 in Bartlett, suggesting that fluoride in drinking water at a concentration of 8 mg/litre might confer some protection against osteoporosis. Later a comparison was made with X-rays of 546 residents of the low-fluoride town of Framingham, Massachusetts (0·04 mg/litre of fluoride). Applying similar criteria to the Texas study, 77 cases of severe osteoporosis were found, significantly more than in Bartlett.[22]

In North Dakota, Bernstein and his colleagues[6] X-rayed the lateral lumbar spine of 300 people from two towns with a high water-fluoride level and 715 people from four other towns with a lower level. Evidence of osteoporosis was appreciably higher in the lower fluoride towns, particularly in women. There were also more cases of vertebral collapse among females in the lower-fluoride communities. This study has been criticised because Grafton, a control town supposedly with a fluoride level in the range 0·15 to 0·30 mg/litre, in fact had a higher level, recorded at different times as 2·8 mg/litre,[26] 0·9 mg/litre[59] and even, in one well some years previously, as 3·6 mg/litre.[1] The choice of the other towns as control has not been criticised. In the case of the high fluoride towns (Mott and Hettinger), levels of 4·0 to 5·8 mg/litre were quoted, as reported in 1962 by Vennes and his colleagues.[60] However, somewhat lower levels have been recorded at other times—of 1·2 in Mott and 2·2 mg/litre in Hettinger[59] though a level of 3·2 mg/litre was found in both towns in 1938.[1] It may be noted, however, that even in the light of these figures the mean level in the 'high-fluoride' towns was apparently still higher than in the control towns, though the difference between the groups is smaller than was initially suggested.

Iskrant[13] suggested that if relative fluoride deficiency encouraged osteoporosis, then the incidence of fractures and their complications (including death) might be higher in low fluoride areas. He therefore examined the mortality rates from falls in urban areas of certain States of the U.S.A. (States with less than five communities with a high natural fluoride level were excluded.) Mortality from falls was lower in naturally high-fluoride (but not in fluoridated) areas than in low-fluoride areas. It was suggested that these findings might be significant and that the absence of a similar effect in fluoridated areas might be due to the shorter duration of exposure to fluoride. Mortality from accidents has also been noted to be lower in high fluoride areas of England and Wales but only in females.[31] In Sweden,[3] however, no significant differences were noted between the incidence of femoral fractures in Eskilstuna (0·8 to 1·2 mg/litre of fluoride) and that in the low-fluoride cities of Malmö (0·2 to 0·4 mg/litre) and Gothenburg (<0·1 mg/litre). No changes in the fracture rates were noted in New York State in Newburgh[22] and Chemung Country[12] after the introduction of fluoridation.

Ansell and Lawrence[4] found that rheumatic complaints, osteoporosis, incapacity due to rheumatism, were all less common in Watford five years after it was fluoridated than in Leigh, a Lancashire town with a negligible amount of fluoride in its water supply. No differences were detected between the two towns in the prevalence of rheumatoid arthritis, osteoarthritis, spondylitis or disc degeneration. Similarly, no difference in the prevalence of osteoporosis was noted between Hartlepool, Co. Durham, which has a fluoride level in its water of over 1 mg/litre and the low-fluoride town of York.[32] A study of the prevalence of musculo-skeletal symptoms in Hartlepool and the nearby low-fluoride town of Teesside has recently been carried out, but no significant differences were detected.[14] Lastly, it may be noted that in spite of an initial impression to the contrary,[17] Ely and his colleagues[8] found that the incidence of Scheuermann's disease and other spinal defects in schoolboys was no higher in areas with a fluoride level of 1·3 to 5·8 mg/litre than in low fluoride areas.

There is no evidence that the prevalence of any musculo-skeletal disorder is increased in areas with fluoride at a concentration of 1 mg per litre in water. Certain evidence suggests that osteoporosis and its complications are less prevalent in high-fluoride areas but it may be noted that some of this evidence refers to levels in excess of 1 mg/litre.

C. SPECIAL CASES

Fluoride is normally cleared from the blood by excretion in the urine and deposition in bone, but in renal failure retention occurs and may cause increased deposition in bone. Patients with chronic renal failure who are dialysed with fluoridated water have an additional load of fluoride from the dialysate,[40,57] and raised fluoride levels in plasma and bone have been reported.[11,19,47,57] Posen raised the possibility that the high prevalence of renal osteodystrophy in the Ottawa haemodialysis centre might be due to the use of fluoridated dialysis fluid.[29] This suggestion was encouraged by the observation that dialysis patients in Newcastle-upon-Tyne and Iowa where the water was also fluoridated tended to develop severe bone disease which was resistant to treatment with vitamin D.[48] In contrast, Montreal, Rochester and the Fulham Hospital, London, where the water was not fluoridated, had a much lower incidence of bone disease.[47] Furthermore, the bone disease in some patients in both Ottawa and Newcastle improved following the use of dialysate fluid made from distilled water.[18] Since the routine use of deionised dialysis fluid in Ottawa was introduced in 1968, none of the 48 patients treated has developed symptomatic osteodystrophy over periods of up to 37 months.[38,39]

However, further work has not supported a simple causative link. Thus, Birmingham, where the water is fluoridated, has little renal osteodystrophy while the Johannesburg dialysis centre, whose water supplies are low in fluoride, has severe bone disease. Moreover, Newcastle had severe bone disease before the introduction of fluoridation in 1968. To investigate this question further, a detailed comparison was made[48] of 94 dialysis patients in Newcastle, and 42 similar patients in Birmingham, which had been fluoridated for four years longer. However, in spite of the fact that the Birmingham patients had been dialysed for twice as long a period as those in Newcastle, osteomalacia, fractures and bone symptoms were all commoner in Newcastle. Because some patients in each centre had also received home dialysis in low-fluoride areas, a further comparison was restricted to those who had only received a high-fluoride dialysate. The findings, however, were the same as before. No differences that might be relevant could be detected in characteristics of the patients or in the medical practices in the centres. Osteoporosis appears to progress more rapidly in Newcastle but there is rather less osteitis fibrosa there than in Birmingham. In neither centre could differences in bone disease be detected between patients living in high and those in low-fluoride areas. It was concluded that some local factor in the water was probably responsible for the severity of the bone disease in Newcastle.

Similarly, in Toronto, a double blind study of the effect of fluoride in dialysis fluid on the progression of osteomalacia was carried out in 11 patients by Oreopoulos and his colleagues[35] with 9 other patients receiving chloride in their dialysate as controls. The average period of observation of the 20 patients in this study was 20·6 months. No difference between the groups was noted in the progression of osteomalacia but significantly more patients in the fluoride group developed radiological evidence of osteosclerosis. It was suggested that this might be beneficial, particularly in those haemodialysis centres that have reported osteoporosis as the main complication.[34,37]

However, in Montreal[7] the degree of progression of bone disease in 34 patients whose dialysate was not fluoridated was recently compared over a period of 46 months with that in 7 patients who used fluoridated dialysis fluid. Of these latter patients, 4 developed severe osteomalacia as opposed to none of the 34 in the non-fluoridated group, suggesting that fluoride in the dialysate, perhaps in conjuction with other substances, promoted the progression of osteomalacia.

In the U.S.A., Juncos and Donadio[16] have reported two asymptomatic patients aged 18 and 17 with pyelonephritis and probable renal dysgenesis who had radiological evidence of skeletal fluorosis. The latter was attributed to fluoride retention caused by reduced excretion of fluoride by the damaged kidneys and also by the polydipsia, which resulted in a large intake of water containing $1 \cdot 7$ mg/litre of fluoride. Asymptomatic osteosclerosis has also been reported in two patients who died from chronic renal failure due, in one case, to chronic pyelonephritis and acute pyelonephritis, and in the other to hydronephrosis caused by a congenital malformation. These occurred in areas of Texas[25] and Argentina,[28] with water-fluoride levels of $4 \cdot 4$ to $5 \cdot 7$ mg/litre and over 2 mg/litre respectively. A case of skeletal fluorosis with radiculomyelopathy has been reported from Texas,[43] U.S.A., in a patient with an excessive intake of water (of uncertain aetiology) who drank water containing fluoride at levels of $2 \cdot 3$ to $3 \cdot 5$ mg/litre for 43 years. No such cases have been reported from areas with fluoride levels of 1 mg/litre.·

No symptomatic cases have been reported of skeletal fluorosis in dialysis patients who have been dialysed with fluid containing 1 mg/litre. Although it is far from consistent, there is evidence that fluoride at a level of 1 mg/litre may contribute to the progression of osteomalacia in patients on long-term haemodialysis. The increasing use of deionisers in such cases therefore seems prudent.

REFERENCES

1. Abbott, G. A. and Voedisch, F. W. (1938). The Municipal Ground Water Supplies of North Dakota. *North Dakota Geological Survey Bulletin 11.*
2. Agate, J. N., Bell, G. H., Boddie, G. F., Bowler, R. G., Buckell, M., Cheeseman, E. A., Douglas, T. H. J., Druett, H. A., Garrad, J., Hunter, D., Perry, K. M. A., Richardson, J. D. and Weir, J. B. deV. (1949). United Kingdom Medical Research Council Memorandum No. 22. *Industrial Fluorosis.* London: H.M.S.O.
3. Alffram, P. A., Hernborg, J. and Nilsson, B. E. R. (1969). *Acta Orthop. Scand.,* **40,** 137.
4. Ansell, B. M. and Lawrence, J. S. (1966). *Ann. rheum. Dis.,* **25,** 67.
5. Azar, H. A., Nucho, C. K., Bayyuk, S. I. and Bayyuk W. B. (1961). *Ann. intern. Med.,* **55,** 193.

6. Bernstein, D. S., Sadowsky, N., Hegsted, D. M., Guri, C. D. and Stave, F. J. (1966). *J. Am. Med. Ass.*, **198**, 499.

7. Cordy, P. E., Gagnon, R., Taves, D. R. and Kaye, M. (1974). *Trans. Amer. Soc. Artif. Int. Organs*, **10**, 197.

8. Eley, A. J., Kemp, F. H., Kerley, P. J. and Berry, W. T. C. (1957). *Lancet*, **ii**, 712.

9. Faccini, J. M. (1969). *Calc. Tiss. Res.*, **3**, 1.

10. Feil, A. (1930). *Paris méd.*, **2**, 242.

11. Fournier, A. E. *et al.* (1971). *J. Clin. Invest.*, **50**, 592.

12. Goggin, J. E. *et al.* (1965). *Pub. Hlth. Rep. (Wash.)*, **80**, 1005.

13. Iskrant, A. P. (1968). *Am. J. Pub. Hlth.*, **58**, 485.

14. Jenkins, G. N. (1975). Unpublished data. Personal communication (July).

15. Jolly, S. S., Singh, B. M. and Mathur, O. C. (1969). *Am. J. Med.*, **47**, 553.

16. Juncos, L. I. and Donadio, J. V. (1972). *J. Am. Med. Ass.*, **222**, 783.

17. Kemp, F. H., Murray, M. M. and Wilson, D. C. (1942). *Lancet*, **ii**, 93.

18. Kerr, D. N. S. (1973). *Proc. E.D.T.A.*, 109.

19. Kim, D. *et al.* (1970). *Trans. Amer. Soc. for Artificial Internal Organs*, **16**, 474.

20. Knizhnikov, V. A. (1958). *Gig. & Sanit.*, **23**, No. 8, 18.

21. Krishnamachari, K. A. V. R. and Krisnaswamy, K. (1973). *Lancet*, **ii**, 877.

22. Korns, R. F. (1969). *Pub. Hlth. Rep. (Wash.)*, **84**, 815.

23. Leone, N. C., Stevenson, C. A., Hilbish, T. F. and Sosman, M. C. (1955). *Amer. J. Roentgen*, **74**, 874.

24. Leone, N. C., Stevenson, C. A., Besse, B., Hawes, L. E. and Dawber, T. R. (1960). *A.M.A. Arch. Ind. Hlth.*, **21**, 326.

25. Linsman, J. F., Crawford, A. and McMurray, D. C. (1943). *Radiology*, **40**, 474. (Also correction in 1943 *Radiology*, **41**, 497).

26. Lohr, E. W. and Love, S. K. (1954). *The Industrial Utility of Public Water supplies in the U.S., 1952.* Part 2. States West of the Mississippi River. Geological Survey Water Supply Paper 1300. Govt. Printing Office, Wash. D.C.

27. McClure, F. J. (1946). In *Dental Caries and Fluorine*. A.A.A.S. Wash. 94.

28. Mascheroni, H. A., Munoz, J. M. and Reussi, C. (1939). *Revista de Sociedad Argentine Biol.*, **15**, 417.

29. Medical Tribune (1969). *Editorial*, **4**, No. 6, 1.

30. Møller, P. F. and Gudjonsson, S. V. (1932). *Acta radiol.*, **13**, 269.

31. Nixon, J. M. and Carpenter, R. G. (1974). *Lancet*, **ii**, 1068.

32. Nordin, B. E. C. (1970). *Osteoporosis* (Ed. Barzel, US.). New York: Grune and Stratton.

33. Ockerse, T. (1941). *South Afr. med. J.*, **15**, 261.

34. Oreopoulos, D. G. *et al.* (1973). *Abstracts of Tenth Congress of European Dialysis & Transplant Association*, 173.

35. Oreopoulos, D. G., Taves, D. R., Ravinovich, S., Meema, H. E., Murray, T., Fenton, S. S. and deVeber, G. A. (1974). *Trans. Amer. Soc. Artif. Int. Organs*, **20**, 203.

36. Pandit, C. G., Raghavachari, T. N. S., Subba Rao, D. and Krishnamurti, V. (1940). *Ind. J. med. Res.*, **28**, 533.

37. Parsons, V., Choudhury, A. A., Wass, J. A. H. and Vernon, A. (1975). *British Medical Journal*, 1, 128.
38. Posen, G. A., Gray, D., Siddiqui, J. Y., Hill, A. V. L., Simpson, W., Ellis, H. and Kerr, D. N. S. (1972). *Q. J. med. N. S.*, 41, No. 164, 534.
39. Posen, G. A., Marier, J. R. and Jaworski, Z. F. (1971). *Fluoride*, 4, 114.
40. Prosser, D. I., Parsons, V., Davies, C. and Goode, G. C. (1970). *Proceedings of the European Dialysis and Transplant Assoc.*, 7, 103.
41. Raghavachari, T. N. S. and Venkataramanan, K. (1940). *Ind. J. med. Res.*, 28, 517.
42. Roholm, K. (1937). *Fluorine Intoxication*. London: Lewis.
43. Sauerbrunn, B. J. L., Ryan, C. M. and Shaw, J. F. (1965). *Ann. intern. Med.*, 63, 1074.
44. Shortt, H. E., McRobert, G. R., Barnard, T. W. and Mannadi Nayar, A. S. (1937). *Indian jour. med. Res.*, 25, 553.
45. Shortt, H. E., Pandit, C. G. and Raghavachari, T. N. S. (1937). *Ind. med. Gazette*, 396.
46. Siddiqui, A. H. (1955). *Brit. med. J.*, Dec. 10, 1048.
47. Siddiqui, J. Y. *et al.* (1970). *Proc. E.D.T.A.*, 7, 110.
48. Siddiqui, J. Y., Simpton, S. W., Ellis, H. E., Kerr, D. N. S., Appleton, D. R., Robinson, B. H., Hawkins, J. B., Robertson, P. W. and Taves, D. R. (1971). *Proc. E.D.T.A.*, 8, 149.
49. Singh, A. and Jolly, S. S. (1961). *Quart. J. Med.*, 41, 357.
50. Singh, A., Jolly, S. S. and Bansal, B. C. (1961). *Lancet*, i, 197.
51. Singh, A., Jolly, S. S., Devi, P., Bansal, B. C. and Singh, S. S. (1962). *Ind. jour. med. Res.*, 50, 387.
52. Singh, A. (1964). In *The Toxicology of Fluorine* (ed. T. Gordonoff). Basel: Schwabe.
53. Singh, A. (1967). *Endemic Fluorosis*. Delhi: Indian Council of Med. Res.
54. Smith, F. A. and Hodge, H. C. (1959). In *Fluorine and dental health* (Muhler, J. C. and Hine, M. K., eds.). Bloomington: Indiana Univ. Press.
55. Stevenson, C. A. and Watson, A. R. (1960). *A.M.A. Arch. industr. Hlth.*, 21, p. 340.
56. Steyn, D. G. (1963). *Pub. Health. Johannesburg*, 63, 34.
57. Taves, D. R., Terry, R., Smith, F. A. and Gardner, D. E. (1965). *Arch. int. Med.*, 115, 167.
58. Teotia, M., Teotia, S. P. S. and Kunwar, K. B. (1971). *Arch. Dis. in Childhood*, 46, 686.
59. U.S. Pub. Hlth. Serv. (1959). *Natural Fluoride Content of Communal Water Supplies in the U.S.* Pub. No. 655.
60. Vennes, J. W., Kwako, J. E., Swenson, D. D. and Peterson, J. K. (1962). *The Journal-Lancet*, 82, 288. Grand Forks and Bismark, North Dakota.
61. Webb-Peploe, M. M. and Bradley, W. G. (1966). *J. Neurol. Neurosurg. Psychiat.*, 29, 577.
62. Weidmann, S. M., Weatherell, J. A. and Jackson, D. (1963). *Proc. nutr. Soc.*, 22, 105.
63. W.H.O. (1970). *Fluorides and Human Health*. Geneva.

64. Zipkin, I., Lee, W. A. and Leone, N. C. (1958). *Proc. Soc. exp. Biol. (N.Y.)*, 97, 650.
65. Zipkin, I., McClure, F. J., Leone, N. C. and Lee, W. A. (1958). *Public Health Reports*, 73, 732.

8 Renal Disorders

It has been suggested that consumption of water containing 1 mg per litre of fluoride might cause or aggravate renal damage. Fluoride is mainly excreted by the kidneys and, in certain cases of renal failure, raised concentrations of fluoride have been recorded in the plasma. (*See* chapter 7(C).) However, studies in areas with fluoride levels in water of 1 mg/litre, and often over, have failed to produce evidence of renal damage by fluoride. There were no more cases of urinary abnormalities in the fluoridated town of Newburgh, New York State, than in the nearby control town of Kingston,[11] or even in areas with fluoride levels of up to 8 mg/litre than in areas with a level of 0·5 mg/litre.[6,7,8]

In his classic study of industrial fluorosis Roholm[10] found only two cryolite workers with any evidence of renal disorder and considered this was doubtfully related to their work. A few cases of impaired renal function and albuminuria have been reported in patients with endemic fluorosis in tropical areas[4,12] but no evidence was presented to suggest that fluoride was the cause. Jolly, Singh and their colleagues in their extensive studies of endemic fluorosis in India[5] have found no evidence of fluoride causing renal dysfunction and observed that they had seen only one patient with fluorosis who also had a renal calculus.[13] It may also be noted that in patients with osteoporosis or Paget's disease treated for long periods with 20 mg or more of sodium fluoride daily there was no significant deterioration in blood urea levels, or in creatinine clearance.[14]

In a study by Hagan and his colleagues[2] of mortality in 32 pairs of U.S. cities with contrasting fluoride levels in their water supplies, mortality from nephritis was higher in 13 and lower in 19 of the 'high-fluoride' than in the 'low-fluoride' cities chosen as controls. Renal diseases were among the causes of death examined by Heasman and Martin[3] in relation to fluoride in England and Wales over the period 1950–59. No significant difference in mortality from renal disease was found between high and low fluoride areas in England and Wales, although when high fluoride areas of Northern England were considered separately, a higher mortality was found than in the corresponding control areas. However, not only was this effect not apparent in Southern England but it was noteworthy that mortality in the Northern high fluoride areas was *lower* than the national or regional averages. This suggests that the differences were due to the chance selection of control towns in the North with unusually low mortality rates from renal disorders. In a U.S. autopsy[1] study no correlation was found between the prevalence or severity of renal disease and the duration of exposure to water containing 2·5 mg/litre of fluoride. Leone and his colleagues[6] noted that an elderly woman who had drunk water with a fluoride level of 4 to 8 mg/litre for 84 years showed no evidence of renal damage at post mortem.

Polyuric renal insufficiency may result from the use in anaesthesia of high concentrations of methoxyflurane and it has been suggested that this is due to inorganic fluoride, one of its breakdown products. However, not only is the contribution by oxalic acid, the other important metabolic product, uncertain but it is not known if molecular methoxyflurane is involved. It is relevant that the concentrations of fluoride produced by this anaesthetic agent are many times higher than those in people who drink water containing 1 mg of fluoride per litre.[9]

There is no evidence that the incidence or mortality of any renal disorder is increased by fluoride in water at a concentration of 1 mg/litre. The question of fluoride in relation to bone disease in patients with chronic renal failure is discussed in Chapter 7(C).

REFERENCES

1. Geever, E. F., Leone, N. C., Geiser, P. and Lieberman, J. (1958). *J. Amer. Dent. Ass.*, **56**, 499.

2. Hagan, T. L., Pasternack, M. and Scholz, G. C. (1954). *Pub. Hlth. Rep.*, **69**, 450.
3. Heasman, M. A. and Martin, A. E. (1962). *Mon. Bull. Min. Hlth.*, **21**, 150.
4. Kumar, S. P. and Kemp Harper, R. A. (1963). *Brit. J. Radiology*, **36**, 497.
5. Jolly, S. S., Singh, B. M. and Mathur, O. C. (1969). *Am. Jnl. Med.*, **47**, 553.
6. Leone, N. C., Shimkin, M. B., Arnold, F. A., Stevenson, C. A., Zimmerman, E. R., Geiser, P. A. and Lieberman, J. E. (1954). *Pub. Hlth. Rep. (Wash.)*, **69**, 925.
7. Leone, N. C., Stevenson, C. A., Hilbish, T. F. and Sosman, M. C. (1955). *Amer. J. Roentgen.*, **74**, 874.
8. McClure, F. J. (1946). Nondental physiological effects of trace quantities of fluorine. In *Dental Caries and Fluorine* (Ed., Moulton, F. R.). Washington, D.C.: Am. Assn. Adv. Sci., p. 74.
9. Mazze, R. I., Trudell, J. R. and Cousins, M. J. (1971). *Anesthesiology*, **35**, 247. [See also *British Medical Journal* Editorial (1972) **2**, 239].
10. Roholm, K. (1937). *Fluorine Intoxication*. London: Lewis.
11. Schlesinger, E. R., Overton, D. E. and Chase, H. C. (1956). *J. Am. Med. Assoc.*, **160**, 21.
12. Siddiqui, A. H. (1955). *Brit. med. J.*, **2**, 1408.
13. Singh, A., Vazirani, S. J., Jolly, S. S. and Bansal, B. C. (1962). *Postgrad. med. J.*, **38**, 150.
14. Taylor, W. II. (1975). Unpublished data. Personal communication (July).

9   Congenital Malformations

MONGOLISM*

The suggestion that fluoride is a cause of mongolism derives from two studies by Rapaport in the U.S.A.[18,19,20] In the first[18] of these, information was obtained about the cases of mongolism registered in institutions in five States. The numbers of cases, grouped according to the town of birth recorded on the birth certificates, were then related to the populations of these towns, and the 'prevalence rates' so obtained compared with the fluoride content of the drinking water. These prevalence rates were, however, not appropriate for the purpose because the place of birth recorded on the certificate was the place where the delivery occurred, which was often different from the mother's place of residence (also recorded on the certificate but not used). In his second study, Rapaport[20] corrected this defect. Information was obtained about children registered in specialist institutions, or recorded on birth or death certificates in one of the five States (Illinois), as mongols born in the years 1950–1956, and whose mothers had resided before the delivery of the child in a town of between 5,000 and 100,000 inhabitants. The incidence of mongolism was then determined by relating the number of cases to the number of births in the same towns in these seven years. His results are summarised in Table 9.1.

At first sight the correlation between the incidence of mongolism and the fluoride content of the water is impressive, until it is noted that the highest rates are only about half those that are normally reported after intensive case finding while the lowest rates are only about one sixth as high. Intensive investigation shows that the incidence of mongolism is remarkably constant, the rate ranging only

* Mongolism is synonymous with Down's Syndrome.

TABLE 9.1

Incidence of mongolism in Illinois
(after Rapaport, 1959, 1963)

Size of towns	No. of towns	Fluoride in water (mg/l)	Mongols	
			No. of cases	Frequency per 1,000 births
10,000	15	0·0	15	0·24
to	24	0·1–0·2	52	0·39
100,000	17	0·3–0·7	33	0·47
inhabitants	12	1·0–2·6	48	0·72
5,000				
to	—	0·0–0·2	10	0·40
10,000	—	0·3–2·6	19	0·78
inhabitants				

from 1·15 to 1·92 per 1,000 births, with a median value of 1·5 per 1,000 in Denmark, Great Britain, Switzerland and the U.S.A.[2]. Rates at this level are universally accepted by paediatricians as compatible with complete ascertainment. To obtain such a level of ascertainment, however, it is usually necessary to utilise, in addition to the sources used by Rapaport, a range of further information, including the records of community physicians, school doctors, midwives, health visitors, social workers and others concerned with the welfare of mongols. It is difficult to attach any meaning to Rapaport's limited enquiry, missing, as it would, most surviving children with mongolism who were cared for at home. Fortunately, it is not necessary to try to do so, as Berry[2] undertook a similar study in nine English towns, making the sort of intensive enquiries that are needed for complete ascertainment. Berry's figures are summarised in Table 9.2, along with those obtained by other British investigators.

The data given in Table 9.2 provide no evidence that the incidence of mongolism bears any relationship to the fluoride content of the drinking water. The absence of any relationship, is, moreover, confirmed by the experience in Birmingham where fluoridation has been practised since

TABLE 9.2

Incidence of mongolism in England, after Berry[1] (1958, 1962)

Author	Place	Fluoride in water (ppm)	Mongols	
			No. of cases	Frequency per 1,000 live births
Berry, 1962	High Wycombe	⎱ less ⎰	9	1·31
	Reading	⎱ than ⎰	30	1·53
	Tynemouth	⎱ 0·2 ⎰	24	1·90
	Carlisle		19	1·64
	Gateshead		37	1·65
	Stockton		16	1·08
	Slough	0·9	13	1·37
	South Shields	0·7–1·1	33	1·59
	W. Hartlepool	1·9–2·0	16	1·23
	6 towns	less than 0·2	135	1·53
	3 towns	more than 0·7	64	1·42
Malpas[13], 1937	Liverpool	—	18	1·29*
Penrose[16], 1949	U.K.	—	7	1·69
Carter and MacCarthy[3], 1951	London	—	107	1·50*
Pleydell[17], 1957	Rural Northants	—	86	1·63*

* Live and still births.

1964. Birmingham is one of the few places in Britain where intensive efforts have been made to secure complete ascertainment of all congenital abnormalities. The annual incidence rates of mongolism from 1960 to 1971 are shown in Table 9.3.[21] Adjustment[3] of these rates to take account of changes in the age at parturition of women in the Birmingham region confirmed that the incidence had not risen since 1964. There is, thus, no indication of any deleterious effect of fluoridation. Similarly, a recent survey in Hartlepool, County Durham, where recorded fluoride levels have always been high (approx. 1·5 to 2·0 mg/litre), has shown that the incidence of mongolism (1967–1974) is in the usual range (1·6 per 1,000 births).[14]

TABLE 9.3

*Incidence of mongolism in Birmingham
1960–1971. (Record[21], 1974)*

Year	Frequency of mongolism per 1,000 live births
1960	1·23
1961	1·77
1962	1·41
1963	1·80
1964*	1·62
1965	1·55
1966	1·59
1967	1·73
1968	1·51
1969	1·04
1970	1·41
1971	1·11

* Start of fluoridation.

This conclusion is further supported by a recent study[15] of the incidence of mongolism in Massachusetts in relation to water fluoride levels. This study utilised the data collected after an intensive case-finding effort on the 1,469 mongol children born to residents of this State in the years 1950–1966.[5] The mean incidence of this disorder was 1·53 per

1,000 total births in the 30 fluoridated towns compared to 1·46 in these towns before fluoridation, and 1·34 in 321 low fluoride communities. A further examination of the incidence of mongolism in these 30 towns was restricted to the 3-year period before fluoridation and the 3-year period after fluoridation was introduced. This produced rates of 1·46 and 1·53 per 1,000 births respectively, a small difference which was mainly attributable to a slight secular increase in the incidence of the disorder throughout the state.

Reports sometimes mentioned in this connection of fluoride causing chromosomal damage in animal experiments[7,9] refer to high concentrations far in excess of those encountered in human tissues in areas with high water–fluoride levels.

OTHER CONGENITAL MALFORMATIONS

Heasman and Martin[8] found mortality in England and Wales from all congenital malformations combined was slightly lower in high than in low fluoride areas, the ratio of the figures for high and low being 0·91. The only specific malformation apart from mongolism that has been suggested as being possibly linked with fluoride is anencephaly. The intake of a large number of dietary items has been examined in relation to perinatal mortality rates for anencephaly.[10] Tea was among several items for which there was a positive correlation, but other food such as cured meats, white bread, canned peas and ice-cream showed stronger correlations. In another study,[6] enquiry of parents of affected children suggested that the prevalence of anencephaly was higher in mothers who drank tea. Because tea has a relatively high content of fluoride it has been suggested that there might be a link between fluoride and anencephaly.[22] There was, however, no increase in prevalence with increase in the consumption of tea above three cups a day. Examination of the prevalence rates for anencephaly in county boroughs with high or high-medium levels of fluoride in their water supplies shows that they have *lower* rates than adjoining county boroughs with lower fluoride levels (Table 9.4). Similarly, in Dublin, which was fluoridated in 1964, the incidence of anencephaly has dropped from 4.37 per 1,000 births in 1953 to 3.3 in 1973.[4]

TABLE 9.4

Mean perinatal mortality rates per 1,000 total births for anencephaly 1963–67 in certain county boroughs (CB)
(Lowe, Roberts and Lloyd (1971)[11] and Lowe (1975)[12])

CBs with high or high-medium water-fluoride levels	CBs in same region with low water-fluoride levels
Birmingham (fluoridated 1964) . . 1·61 (179)	Stoke on Trent . . . . . . 2·79 (63) Walsall . . . . 1·70 (24) West Bromwich . . . . . . 1·28 (15) Coventry . . . . 1·71 (57) Wolverhampton . . . . . 1·86 (37) Mean . . . . 1·93 (196)
West Hartlepool* . . . . 1·54 (10) South Shields† . . . . . . 2·41 (23) Sunderland† . . . . 1·81 (34) Mean . . . . 1·92 (67)	Darlington . . . . 1·51 (11) Gateshead . . . . 3·44 (33) Middlesbrough . . . . . . 2·04 (23) Newcastle-u-Tyne . . . 1·85 (41) York . . . . 1·71 (15) Mean . . . . 2·08 (133)

(Numbers in parentheses refer to the number of cases).

* Fluoride level over 1 p.p.m. Figures for this CB do not include 1967 when a boundary change occurred.

† Fluoride level 0·5–0·7 p.p.m.

REFERENCES

1. Berry, W. T. C. (1958). *Amer. J. Ment. Defic.*, **62**, 623–636.
2. Berry, W. T. C. (1962). *Med. Off.*, **108**, 204–2
3. Carter, C. and MacCarthy, D. (1951). *Brit. J. of Soc. Med.*, **5**, 83–90.
4. Coffey, V. P. (1974). *J. Irish med. Assoc.*, **67**, 553.
5. Fabia, J. and Drolette, M. (1967). *Pediatrics*, **45**, 60.
6. Fedrick, J. (1974). *Proc. roy. soc. Med.*, **67**, 356.
7. Gileva, E. A., Plotko, E. G. and Gatiyatullina, E. E. (1972). *Gig. Sanit.*, **37**, 9.
8. Heasman, M. A. and Martin, A. E. (1962). *Mon. bull. Min. Hlth.*, **21**, 250.
9. Jagiello, G. and Lin, J. S. (1974). *Arch. Environ. Health*, 29, 230.
10. Knox, E. G. (1972). *Brit. J. Prev. Soc. Med.*, **26**, 219.
11. Lowe, C. R., Roberts, C. J. and Lloyd, S. (1971). *Brit. med. J.*, **2**, 357.

12. Lowe, C. R. (1975). Personal communication.
13. Malpas, P. (1937). *J. Obst. Gynaec. and Brit. Emp.*, **44**, 434–454.
14. Milligan, H. C. (1975). Personal communication (June).
15. Needleman, H. L., Pueschel, S. M. and Rothman, K. J. (1974). *New England J. Med.*, **291**, 821.
16. Penrose, L. S. (1949). *J. ment. Sci.*, **95**, 685.
17. Pleydell, M. J. (1957). *Lancet*, **i**, 1314.
18. Rapaport, I. (1956). *Bull. Acad. Nat. Med.*, (Paris), **140**, 529–531.
19. Rapaport, I. (1959). *Bull. Acad. Nat. Med.*, (Paris), **143**, 367–370.
20. Rapaport, I. (1963). *Rev. Stomat* (Paris), **46**, 207–218.
21. Record, R. G. (1974). Unpublished data. Personal communication.
22. Sinclair, H. (1973). *Lancet*, **ii**, 962.

10 Cancer

It has sometimes been said that fluoride can cause cancer, and reference has been made to certain experiments on animals or cell cultures. Taylor and Taylor[17] reported that fluoride accelerated the growth of neoplastic cell cultures in eggs and of transplanted tumours in mice whether injected or given in drinking water at a level of 1 mg/litre. This work has not been confirmed. Fluoride did not stimulate growth of the Walker rat sarcoma in experimental animals,[4] while the growth of a transplanted sarcoma in mice and guinea-pig was inhibited by injections of sodium fluoride.[5] Berry and Trillwood[2] reported that fluoride at a concentration of $0 \cdot 1$ mg/litre inhibited the growth of certain cell cultures including a HeLa cell culture (originally from a cervix cancer). This work has been repeated but effects were detected only at concentrations of fluoride in the culture media far higher than that possible in human serum.[1,11] No carcinogenic or tumour accelerating effects have been detected following toxicity studies of fluoride compounds in experimental animals.[3,9,16]

Heasman and Martin[8] in their study of mortality in relation to fluoride levels in water found a higher mortality from cancer of the stomach in certain high-fluoride areas in northern England. No such association was observed in Southern England, and two areas in the North with a common high-fluoride water supply had markedly different mortality rates from stomach cancer. They therefore concluded that the association observed in the North was coincidental, a conclusion supported by a recent study.[12] It may also be noted that Stocks[15] found no excess of stomach cancer in heavy tea drinkers, a group with a relatively high fluoride intake.

The incidence of cancer of the stomach and of several other organs has recently been investigated in relation to fluoride.[10] By means of the national cancer registration scheme it was possible to compare cancer incidence in areas with contrasting levels of fluoride in their water. This approach has the advantage that the collection of the data was entirely independent of any hypothesis connected with fluoride.

For each local authority district with a fluoride level in its water supply of 1 mg/litre or over ('high'), one or more nearby districts of similar size were selected with a fluoride level of 0·2 mg/litre or less (low). For areas with a mean fluoride level in the range 0·5 to 0·99 mg/litre (high-medium), control areas were chosen with fluoride levels of 0·1 mg/litre or less (very low). Numbers of cancers of each site were aggregated by age group and sex into these four categories according to whether the water-fluoride level in the area of residence was 'high', 'high-medium', 'low', or 'very low', as defined above. For each site of cancer, the numbers of cancers observed in each category of area were then compared within each age

TABLE 10.1

Ratios of observed to expected numbers of cancers in certain organs in areas with different levels of fluoride (F) in water

Site of cancer	High F (>1)		High-medium F (0·5–0·99)		Low F (<0·2)		Very low F (<0·1)	
Thyroid*	1·05	(45)	0·79	(54)	1·27	(57)	1·02	(84)
Kidney*	1·01	(129)	1·00	(198)	1·02	(131)	0·98	(233)
Stomach	0·88	(375)	1·15	(733)	0·90	(327)	1·05	(815)
Oesophagus	0·87	(73)	1·02	(131)	0·87	(73)	1·13	(177)
Colon	0·96	(386)	1·03	(613)	0·99	(385)	1·00	(719)
Rectum	0·93	(273)	1·11	(486)	0·94	(264)	0·99	(519)
Bladder*	1·00	(430)	0·96	(632)	1·06	(444)	1·00	(786)
Bone**	1·00	(18)	1·06	(30)	1·02	(19)	0·94	(31)
Breast	0·92	(567)	1·06	(999)	1·08	(650)	0·97	(1105)
Total population	482,398		779,054		510,045		896,625	

In parentheses are the total numbers of cancers observed.

The above data relate to the years 1961, 1963, 1965 and 1967, except for the sites marked *, which cover the whole period 1961–1968, and **, which refer to 1962, 1964, 1966 and 1968. The same applies to Table 10.2.

group and sex with the numbers expected if the incidence of the disease had been uniform throughout the four areas. Table 10.1 shows, for cancers of the thyroid, kidney, stomach, oesophagus, colon, rectum, bladder, bone, and breast, the numbers recorded in each of the four sets of districts grouped by water fluoride level together with the ratios of observed to expected numbers. It will be seen that there is no tendency for the ratio for any cancer to be greater in the high-fluoride areas than in the low-fluoride areas.

A comparison was also made between cancer incidence in the three areas of England and Wales that were fluoridated before 1968 and neighbouring low fluoride areas. These fluoridated areas were Angelsey, Watford and Birmingham (together with Solihull), the schemes having been introduced in 1955, 1956, and 1964 respectively. The results of this analysis are summarised in Table 10.2 and indicate no significant excess of cancer in fluoridated areas as compared with nearby unfluoridated areas. Data were also examined on the incidence of cancer of the thyroid, kidney, and bladder in fluoridated and low-fluoride areas in other countries; namely, the U.S.A. (New York State and Connecticut), New Zealand and Holland. In none of these areas was there any tendency for the incidence of these cancers to be higher in the fluoridated areas than in the low-fluoride areas; and, if anything, the opposite was the case.[10]

The lack of any relationship between fluoride and cancer incidence is in keeping with mortality studies of this question.[6,7,8,12,18] An examination of mortality in 32 pairs of U.S. cities with contrasting water-fluoride levels showed an exact balance, with 16 high-fluoride cities showing a higher mortality from cancer than their control cities, but 16 others showing a lower rate.[6,7] As in other countries many urban areas in the U.S.A. have cancer mortality rates above the national average and because some of these areas are fluoridated, the question of a link between fluoride and cancer has been raised.[18] More detailed studies by the National Cancer Institute including an examination of rates before and after fluoridation have produced no evidence to support this suggestion.[3,18] Lastly, and contrary to a recent implication,[14]

TABLE 10.2

Ratio of observed to expected numbers of cancers in fluoridated and non-fluoridated areas with low levels of natural fluoride in the water.

Site of cancer	Fluoridated 1·0 mg		Non-fluoridated (<0·15)	
Thyroid*	1·09	(100)	0·91	(81)
Kidney*	1·04	(223)	0·96	(201)
Stomach	0·98	(678)	1·02	(684)
Oesophagus	0·99	(141)	1·01	(140)
Colon	1·03	(634)	0·97	(572)
Rectum	1·06	(480)	0·94	(416)
Bladder*	1·04	(730)	0·96	(658)
Bone**	0·88	(20)	1·18	(28)
Breast	1·03	(1099)	0·97	(986)
Total population (mean of 1961 and 1971)	1,295,212		1,304,676	

In parentheses are the total number of cancers observed.

In the case of data for Birmingham and Solihull and their control towns, the data relate only to the period 1965–68, since fluoridation was only introduced in these towns in October 1964. (*See* note below Table 10.1).

mortality from cancer, and from leukaemia in particular, has not increased in Birmingham since it was fluoridated relative to low fluoride parts of the same region; slight increases have occurred in both, but these were greater in the low fluoride areas.[13]

There is no evidence that fluoride increases the incidence or mortality of cancer in any organ.

REFERENCES

1. Armstrong, W. D., Pollock, M. E. and Singer, L. (1965). *Brit. Med. J.*, **5441**, 1435.
2. Berry; R. J. and Trillwood, W. (1963). *Brit. med. J.*, **5364**, 1064.
3. Berg, J. (1972). Personal communication of U.S. National Cancer Institute (1972) Statement on Alleged Association of Fluorides with Cancer.
4. Finerty, J. C. and Grace, J. D. (1952). *Texas Reports on Biology and Medicine*, **10**, 501.

5. Fleming, H. S. (1953). *J. dent. Res.*, **32**, 646.
6. Hagan, T. L. (1969). In *Fluorine and Dental Health* (ed. Muhler, J. C. and Hine, M. K.). Bloomington: Indiana University Press.
7. Hagan, T. L., Pasternack, M. and Scholz, G. C. (1954). *Pub. Hlth. Rep.*, **69**, 450.
8. Heasman, M. A. and Martin, A. E. (1964). *Mon. Bull. Minist. Hlth*, **21**, 150.
9. Kanisawa, M. and Schroeder, H. A. (1969). *Cancer Research.* **29**, 892.
10. Kinlen, L. J. (1975). *Brit. dent. J.*, **138**, 221.
11. Nias, A. H. W. (1965). *Brit. med. J.*, **5451**, 1672.
12. Nixon, J. M. and Carpenter, R. G. (1974). *Lancet*, **ii**, 1068.
13. Registrar General, Statistical Reviews of England and Wales, Pt. 1 Mortality. 1960–1972.
14. Schatz, A. and Schatz, V. (1972). *Compost Science*, **13**, 27.
15. Stocks, P. (1958). *Supplement to 35th Report Brit. Emp. Cancer Campaign (Pt. 2)*.
16. Tannenbaum, A. and Silverstone, H. (1953). *Cancer Research*, **13**, 460.
17. Taylor, A. and Taylor, N. C. (1965). *Proc. Soc. Exp. Biol. Med. (N.Y.)*, **119**, 252.
18. U.S. Department of Health, Education and Welfare. Statement FL-76: 'National Cancer Institute Rejects Fluoride Scare Report'. Congressional Record—House for July 21, 1975. H.7175–6, and personal communication from Dr. R. N. Hoover of the N.C.I., October, 1975.

11 Other Conditions

A. ALLERGY AND INTOLERANCE

Many different disorders have been regarded by Waldbott[50-52] as evidence of allergy or intolerance to fluoride at a level of 1 mg/litre in drinking water. These disorders include—

oral and gastrointestinal disturbances such as—
> stomatitis, gingivitis, oral ulcers, flatulence, nausea, abdominal pain, vomiting, haematemesis, diarrhoea and constipation;

neurological and muscular symptoms such as—
> headache, paraesthesiae, painful numbness in the extremities, mental deterioration, convulsions, personality change, backache and pain on 'muscular assertion of the ribs' (sic);

urinary tract disorders such as—
> urethritis, cystitis and pyelitis;

skin conditions such as—
> urticaria, skin rashes and brittle nails;

non-specific symptoms such as—
> weakness and exhaustion after sleep.

These symptoms were reported from fluoridated parts of the U.S.A. by 52 people, apparently in response to a published request for details of adverse reactions to drinking fluoridated water. Some of these symptoms were similar to those that 11 patients reported to Waldbott as having improved after fluoridation was discontinued in Saginaw, Michigan.

The fact that these observations were uncontrolled is important since many of these complaints are common in the general population irrespective of the fluoride content of water supplies. The need for caution in assessing such reports is underlined by the fact that complaints attributing digestive and other disorders to fluoridated water have in some cases preceded the actual start of fluoridation.[9,20]

Waldbott later reported that 123 selected allergic patients who were given 15 mg sodium fluoride, 5 developed symptoms after an interval of 5 minutes to 3 hours lasting from 10 hours to 10 days. Of the wide variety of about 20 symptoms and signs listed, only 7 occurred in more than one patient: namely, paraesthesiae, nausea, vomiting, abdominal pain, headache (or migraine), stomatitis and lethargy.

Waldbott has also reported[51] a double-blind experiment involving 48 patients with suspected intolerance to fluoride in water. Each patient was provided with three similar bottles of distilled water, one of which contained 1 mg of sodium fluoride per tablespoon, and instructed to take one table-spoon from a given bottle for a week, and then repeat the procedure with the other bottles in successive weeks. Of these 48 people, 29 reported ill effects such as joint pains and headaches which lasted for up to 10 days, though it was not specifically stated that these ill effects were related to the fluoride solution. No mention was made of any symptoms experienced while taking distilled water alone, such as are commonly noted in trials using a placebo. It may also be noted that distilled water containing 1 mg of sodium fluoride per tablespoon has a distinctive taste.

It may be noted that studies in parts of the U.S.A. where the fluoride concentration is 8 mg/litre have produced no evidence of this syndrome described by Waldbott. Moreover, patients with skeletal fluorosis in India did not report such symptoms.[24,41] Some patients with osteoporosis or Paget's disease treated with 20 to 120 mg of sodium fluoride daily (equivalent to 9 to 55 mg fluoride ion) experience gastro-intestinal symptoms (*see* Chapter 5, page 28).

There is a marked difference in concentration between taking even a single sodium fluoride tablet (9 mg of fluoride ion) with one cup of water and the same amount of fluoride taken in about 2 gallons of water (containing 1 mg/litre over many hours). Investigation of individual cases of supposed intolerance of fluoridated water have failed to confirm fluoride as the cause.[9,11]

As a result of reports of allergy and intolerance the U.S. Public Health Service asked the American Academy of Allergy to evaluate available clinical reports in terms of the main types of allergic response; and also to examine the possibility that certain cases might belong to less well understood types of drug reaction. A statement was later issued by the Academy[21] that there was no evidence of allergy or intolerance to fluorides as used in the fluoridation of community water supplies.[23]

B. THYROID DISORDERS

Because of the chemical similarities between iodine and fluorine there has been much interest in the possible effects of fluoride on thyroid function and, indeed, a century ago it was suggested as a cause of endemic goitre.[32] Since then, the role of iodine in thyroid metabolism has been elucidated and endemic goitre has now virtually disappeared following the provision of iodine supplements such as iodised salt to communities irrespective of their fluoride intake.[7,19,53] Because there seemed some similarity both in England and in the Punjab[4,54] in the distribution of endemic goitre and dental mottling it was suggested that both were caused by fluoride. No account, however, was taken of iodine supply in these reports. Later surveys in England, which remedied this omission, failed to produce evidence of a goitrogenic effect of fluoride.[2,33] Singh and his colleagues[41] have pointed out that the endemic goitre area in the Punjab is in fact distinct from the endemic fluorosis area. Siddiqui[40] examined the thyroid in all 2,000 inhabitants of three villages in an area of endemic fluorosis and found no case of obvious goitre. More instances of slight thyroid enlargement at ages 14 to 17 years were noted in a village with a fluoride level in its water of 10·7 mg/litre than in a village with a level of 5·4 mg/litre but it may be noted that the former had a lower iodine intake in its water supply.

Day and Powell-Jackson[5] reported a lower prevalence of

goitre in Nepal in two villages where the fluoride level averaged 0·12 and 0·19 mg/litre respectively than in two villages with mean levels of 0·24 and 0·26 mg/litre. It has been pointed[16,22] out that details of total iodine intake were not available for any of the communities studied and that the higher fluoride levels are within what is usually regarded as a 'low' range and are far below the levels in endemic fluorosis areas where no such relationship has been noted. Particular attention was paid to the thyroid in the ten-year study of children living in Newburgh after it was fluoridated in comparison with those living in the low-fluoride town of Kingston but no differences were observed.[37] Lastly, it may be noted that the thyroid is not affected in industrial fluorosis.[36]

Fluoride is not concentrated by the thyroid and does not influence its uptake of iodine.[7,12,13,26,29,55] Compared to a control group, no differences were noted in serum thyroxine levels in 36 patients with osteoporosis or Paget's disease who received 20 to 60 mg of sodium fluoride daily for long periods.[49] An improvement in some cases of hyperthyroidism has been reported following prolonged treatment with fluoride but the mechanism of action is not clear.[7,13]

In summary there is no evidence that fluoride is responsible for any disorder of the thyroid.

C. OTHER ENDOCRINE DISORDERS

In the Second World War, Spira asked by questionnaire 1,099 military[43] personnel with mottled teeth whether they suffered from constipation, paraesthesiae, boils, heat rashes, loose shrivelled skin between the toes, loss of hair and brittle nails. On the basis of the proportion reporting, these symptoms were attributed to parathyroid[45,46] disorders caused by fluoride. Mottled teeth were equated with dental fluorosis and no attempt was made to exclude other causes of mottling. Indeed, the counties from which most subjects of this study came did not correspond with known high-fluoride areas.[42] Data were not presented from personnel without mottled teeth. Later,[47] adrenal dysfunction was also invoked to account for lassitude. No investigations were carried out to confirm these supposed endocrine disturbances but other

workers have found normal adrenal and parathyroid function in cases of skeletal fluorosis.

Because fluoride in large doses can impair enzymatic
processes it has sometimes been suggested that it might cause
diabetes by blocking the action of enzymes involved in
glucose metabolism. Roholm[36] recorded that one of the
cryolite workers in his study had diabetes, but found no cases
of unsuspected glycosuria in people at risk of industrial
fluorosis. He did not regard fluoride as a cause of diabetes.
Intensive studies of endemic fluorosis in India have also
produced no evidence of fluoride causing diabetes.[41] Similarly, routine examinations of children in Newburgh in the ten
years after it was fluoridated revealed no more urinary abnormalities than in Kingston, a low-fluoride town.[37]

D. OTHER DISORDERS

(a) CARDIOVASCULAR DISORDERS

Heasman and Martin[18] examined mortality from all major
causes of death in England and Wales in the ten year period
1950–59 in relation to fluoride levels in drinking water. In the
case of coronary disease and other cardiac disorders, a lower
mortality was noted in the high fluoride areas but was considered probably to be due to differences in the hardness of
the water. A recent analysis[34] which took account both of
water hardness and certain socio-economic variables, found
coronary mortality was still lower in high fluoride areas but
this was not statistically significant. The opposite suggestion,
that fluoride increases coronary mortality, has been made
because of an impression that this had happened in Antigo,[21]
a small community of 9,000 in Wisconsin, which was
fluoridated in 1949. However, because of the secular trends
in coronary mortality, there is a particular need for controls
before such deductions are valid. No increase in coronary
mortality was noted after 10 years fluoridation in Newburgh,[37] New York State, or after 15 years in Easton,[3] Pennsylvania, relative to that in low fluoride communities. There
has been no increase in mortality from coronary disease in
Birmingham since it was fluoridated in 1964, relative to low

fluoride parts of its region.[35] Leone and his colleagues[27,28] found slightly fewer cardiovascular abnormalities in a sample of residents in the high-fluoride town of Bartlett (8 mg/litre) than in Cameron a nearby town in Texas (fluoride 0·4 mg/litre). There is no epidemiological evidence to support the suggestion[10] that fluoride increases the prevalence of arteriosclerosis.

Bernstein and his colleagues[1] made the unexpected observation that the prevalence of aortic calcification was lower in areas with high fluoride levels than in other areas where the level, though still appreciable, was somewhat lower. This difference was noted in both sexes but was significant only in males. No explanation could be offered for this finding and Korns[25] found no such effect in his study of Newburgh and Kingston more than ten years after the former was fluoridated.

(b) OPTIC NEURITIS

Optic neuritis has been connected with fluoride because of a single case report,[14] in 1964, of this disorder developing in a 56-year-old man six weeks after starting sodium fluoride therapy (60 mg daily) for severe spinal osteoporosis, and about 6 months after a total parathyroidectomy. No similar cases have been documented although sodium fluoride has since been widely used in the treatment of both osteoporosis and Paget's disease, and optic nerve damage has been looked for specifically.[6,48]

(c) OTOSCLEROSIS

Spira[44] also suggested that fluoride might play a part in otosclerosis because of his impression of histological similarities between the temporal bone in otosclerosis and teeth affected by fluorosis. Also, in a few cases of this disorder, the administration of 30 mg of fluoride daily for two months produced slight changes in the otosclerotic focus.[38] However, tests of auditory nerve function in cases of severe endemic fluorosis in India do not suggest the presence of otosclerosis.[39] In fact, audiometric tests on children indicated a higher prevalence of defective hearing (4·9 per cent) in low-

fluoride areas of Illinois than in nearby areas with higher
fluoride levels (2·8 per cent).[30]

(d) MISCELLANEOUS DISORDERS

A wide variety of disorders in addition to those already discussed have been covered in mortality and morbidity studies
connected with fluoride. Thus, Hagen[17] and his colleagues
found no difference in mortality from heart disease, in-
tracranial lesions or cirrhosis in 16 pairs of U.S. cities with
contrasting fluoride levels in their water supplies of 0·7
mg/litre or more, and under 0·25 mg/litre respectively. All
major causes of death were included by Heasman and Martin[18] in their analysis of mortality in relation to fluoride levels
in water. An excess of deaths from bronchitis and pneumonia
in high-fluoride areas was noted. This was considered unlike-
ly to be caused by fluoride since the effect was reversed in the
case of bronchitis in the southern and northern groups of
towns and in these respiratory conditions, wide geographical
differences are known to be related to social, economic and
other factors. Stillbirths and infant mortality rates were also
examined in this study in high and low fluoride areas and a
significantly lower stillbirth rate was noted in the high-
fluoride towns. Again, this difference was probably due to
chance.

Geever and his colleagues[15] reported on their findings in
over 700 post-mortem examinations on residents of
Colorado Springs, which has a fluoride level in the water of
2·5 mg/litre. No significant differences in any disease could
be detected between those who had lived in this high-fluoride
town for over 20 years, for 5 to 20 years, or for less than 5
years.

At least two studies[27,37] have involved a detailed medical
examination supplemented by special investigation of
residents of high-fluoride areas to detect possibly unsuspected effects of fluoride. Leone and his colleagues[27]
examined a sample of 116 residents of Bartlett (water fluoride
level 8 mg/litre) and 121 residents of Cameron (fluoride level
of 0·4 mg/litre) in 1943, and again in 1953. No differences
were detected except that there was more dental fluorosis in

Bartlett and slightly more cardiovascular abnormalities in Cameron. In New York State, Schlesinger and his colleagues[37] examined a sample of 817 children in Newburgh just before it became fluoridated and 711 children from a control town, Kingston, and a proportion of these at regular intervals over the next ten years. A threefold higher tonsillectomy rate in Newburgh could not be attributed to fluoride since it was present before fluoridation was introduced. No difference was detected with respect to any disorder, other than caries, that could be ascribed to fluoridation.

Singh and his colleagues,[41] in their studies of endemic fluorosis in the Punjab submitted a detailed questionnaire to the inhabitants of an area with a fluoride level of 10 mg/litre in the drinking water. From the answers they received, supplemented by a detailed cardiovascular examination, they could find no significant evidence of anaemia, underdevelopment, or any non-skeletal disorder. Indeed, they observed that the male residents of the area in question were among the tallest and fittest in the country.

Other comparisons made in high and low fluoride areas include absenteeism among school children, anaemia in pregnancy, perforation of peptic ulcers[8] and X-rays of bones of children,[31] but no differences have been detected.

There is no evidence that allergies, thyroid disorders or any of the other conditions mentioned above can be caused by 1 mg/litre of fluoride in drinking water.

REFERENCES

1. Bernstein, D. S., Sadowsky, N., Hegsted, D. M., Guri, C. D. and Stave, F. J. (1966). *J. Am. Med. Ass.*, **198,** 499.
2. Berry, W. T. C. and Whittles, J. H. (1963). *Mon. Bull. Min. Hlth.*, **22,** 50.
3. Bierenbaum, M. L. and Fleischman, A. I. *et al.* (1974). *J. Med. Soc., N.J.*, **71,** 663.
4. Bromehead, C. N., Murray, M. M. and Wilson, D. C. (1943). *Lancet*, **i,** 490.
5. Day, T. K. and Powell-Jackson, P. R. (1972). *Lancet*, **i,** 1135.
6. de Deuxchaisnes, C. N., Krane, S. M. (1964). *Medicine*, **43,** 233.
7. Demole, V. in W.H.O. (1970). *Fluoride and Human Health*, p. 225. Geneva.

8. Department of Health and Social Security (1962). *Rep. on Pub. Hlth. and Med. Subjects No. 105.* London: H.M.S.O.
9. Department of Health and Social Security (1969). *Rep. on Pub. Hlth. and Med. Subjects No. 122.* London: H.M.S.O.
10. Exner, F. B. and Waldbott, G. L. (1957). *The American Fluoridation Experiment.* New York: Devin-Adair.
11. Fremlin, J. H. (1967). *Brit. dent. J.*, **122**, 178.
12. Gabovich, R. D. (1960). *Gig, Tr. prof. Zabolos 2*, 26 quoted in W.H.O. 1970 Fluoride and Human Health, p. 225. Geneva.
13. Galletti, P. M. and Joyet, G. (1958). *J. Clin. Endocrin.*, **18**, 1102.
14. Geall, M. G. and Beilin, L. J. (1964). *Brit. med. J.*, **3**, 355.
15. Geever, E. F., Leone, N. C., Geiser, P. and Lieberman, J. (1958). *J. Am. dent. Ass.*, **56**, 499.
16. Hadjimarkos, D. M. (1972). *Lancet*, **i**, 1343.
17. Hagan, T. L., Pasternack, M. and Scholz, G. C. (1954). *Pub. Hlth. Rep.*, **69**, 450.
18. Heasman, M. A. and Martin, A. E. (1962). *Mon. Bull. Min. Hlth.*, **21**, 250.
19. Held, A.-J. (1967). Quoted in W.H.O. 1970. *Fluoride and Human Health*, p. 260.
20. Hilleboe, H. E. (1956). *J. Am. dent. Ass.*, **52**, 291.
21. Janson, I. and Thomson, H. M. (1974). *Fluoride*, **7**, 62.
22. Jenkins, G. N. and Cooke, J. A. (1972). *Lancet*, **i**, 1293.
23. *Journal of Allergy* (1971). Editorial, **47**, 347.
24. Jolly, S. S., Singh, B. M. and Mathur, O. C. (1969). *Amer. J. Med.*, **47**, 553.
25. Korns, R. F. (1969). *Pub. Hlth. Rep. (Wash.)*, **84**, 815.
26. Korrodi, H. *et al.* (1956). *Helv. Med. Acta*, **23**, 601.
27. Leone, N. C., Shimkin, M. B., Arnold, F. A., Stevenson, C. A., Zimmerman, E. R., Geiser, P. A. and Lieberman, J. E. (1954). *Pub. Hlth. Rep. (Wash.)*, **69**, 925.
28. Leone, N. C., Stevenson, C. A., Hilbish, T. F. and Sosman, M. C. (1955). *Amer. J. Roentgen*, **74**, 874.
29. Levi, J. E. and Silberstein, H. (1955). *J. Lab. and Clin. Med.*, **45**, 348.
30. Lewy, A. (1944). *Arch. Otolaryngol.*, **39**, 152.
31. McCauley, H. B. and McClure, F. J. (1954). *Pub. Hlth. Rep. (Wash.)*, **69**, 671.
32. Mauméné, E. (1866). *C.R. Acad. Sci., Paris*, **62**, 381. Quoted in W.H.O. (1970).
33. Murray, M. M., Ryle, J. A., Simpson, B. W. and Wilson, D. C. (1948). *M.R.C. Memorandum*, No. 18.
34. Nixon, J. M. and Carpenter, R. G. (1974). *Lancet*, **ii**, 1450.
35. *Registrar General's Annual Statistical Review of England and Wales 1959–1973.* London: H.M.S.O.
36. Roholm, K. (1947). *Fluorine Intoxication.* London: Lewis.
37. Schlesinger, E. R., Overton, D. E. and Chase, H. C. (1956). *J. Am. Med. Ass.*, **160**, 21.

38. Shambaugh, G. E., Jr., and Scott, A. (1964). *Arch. Otolaryngol.*, **80**, 263.
39. Siddiqui, A. H. (1955). *Brit. med. J.*, **2**, 1408.
40. Siddiqui, A. H. (1960). *J. Endocrinol.*, **20**, 101.
41. Singh, A., Vazirani, S. J., Jolly, S. S. and Bansal, B. C. (1962). *Postgrad. Med. J.*, **28**, 150.
42. Spira, L. (1942). *Lancet*, **i**, 64.
43. Spira, L. (1942). *Edinburgh med. J.*, **49**, 707.
44. Spira, L. (1943). *J. Laryngology*, **58**, 151.
45. Spira, L. (1943). *Edinburgh med. J.*, **50**, 237.
46. Spira, L. (1943). *J. Hygiene*, **43**, 402.
47. Spira, L. (1953). *The Drama of Fluorine, Arch Enemy of Mankind.* Milwaukee, Wis., Lee Foundation for Nutritional Research. Quoted in W.H.O. (1970).
48. Taylor, W. H. (1970). *Brit. med. J.*, **4**, 304.
49. Taylor, W. H. (1975). Unpublished data. Personal communication (July).
50. Waldbott, G. L. (1955). *Acta. med. Scand.*, **156**, 157.
51. Waldbott, G. L. (1962). *Int. Arch. Allergy and Appl. Immunol.*, **20**, Suppl. 1, 1.
52. Waldbott, G. L. (1963). *Acta. med. Scand.*, **174**, Suppl. 400, 1.
53. Wespi, H. J. (1954). *Praxis*, **43**, 616. Quoted in W.H.O. (1970).
54. Wilson, D. C. (1941). *Lancet*, **i**, 211.
55. World Health Organisation (1970). *Fluorides and Human Health*, Geneva.

I. THE PRESENT STATE OF FLUORIDATION

EXTENT OF CONSUMPTION OF HIGH-FLUORIDE WATER

Fluoridation is practised in more than 30 countries.[43] Table 12.1 summarises the world position in 1969 when schemes serving over 120 million people had been initiated, a population that was estimated to have risen to 150 million by 1971.[34] In England and Wales $4\frac{3}{4}$ million people are currently receiving fluoridated water.[16,17] In addition, about 10 million people in Scandinavia, the United Kingdom and the U.S.A. receive water containing fluoride which is naturally present at a level of at least 0·7 mg/litre[24,40,45] (Table 12.2).

TECHNICAL ASPECTS OF FLUORIDATION

Regular or continuous monitoring of the fluoride concentration in water supplies together with fail-safe devices ensure the maintenance of the fluoride level within narrow limits. In Grand Rapids, Michigan, where fluoridation was first introduced, 99 per cent of about 3,000 tests over a ten-year period were in the range 0·8 to 1·2 mg/litre.[10] In San Francisco, 640 tests in one year showed a variation of only 0·04 mg/litre from the target level.[3] Technical advances have ensured the efficiency of fluoridation plants and any temporary increase in the fluoride level automatically results in the fluoride feeder being turned off.

It could logically be suggested that if the optimal concentration of fluoride in drinking water is approximately 1 mg/litre, the level should be reduced in supplies in which the concentration of fluoride present is higher than it is naturally. It is relevant that, as a result of reorganisation of water supplies, there are no longer any communities in

TABLE 12.1

Fluoridation throughout the world

Country or territory	Year first project started	No. of communities reported to have fluoridated water	Population served
Australia	1956	23	4,159,000
Belgium	1956	1	10,000
Brazil	1953	86	1,500,000
Canada	1945	459	6,063,000
Chile	1953	64	2,401,000
Columbia	1953	6	1,000,000
Czechoslovakia	1958	30	1,380,000
El Salvador	1956	1	
Federal Republic of Germany	1952	1	6,000
Finland	1959	1	60,000
German Democratic Republic	1959	2 or more	310,000*
Hong Kong	1961	11	3,570,000
Ireland	1964	44	1,200,000
Japan	1952	—	—
Kuwait	1968	1	676,000
Malaysia	—	6	3,000,000
Mexico	1960	5	1,750,000
Netherlands	1953	15	3,000,000
New Zealand	1953	29	1,205,000
Panama	1950	8	510,000
Papua and New Guinea	1966	1	38,000
Paraguay	1959	1	135,000
Poland	—	1	500,000
Puerto Rico	1953	59	1,808,000
Romania	—	1	100,000
Ryukyu Island	—	2	740,000
Sarawak	—	—	180,000
Singapore	1958	1	2,000,000
Sweden	1951	2	130,000
Switzerland	1960	3	250,000
United Kingdom	1955	15	2,250,000
U.S.A.	1945	3,827	74,600,000
U.S.S.R.	1960	24	13,000,000
Venezuela	1952	22	60,000

(W.H.O. Chronicle, 1969).   * Forrest, 1967[12].

Britain that are constantly exposed to a fluoride concentration in excess of 2 mg/litre a level at which no harmful effects have been demonstrated in a temperate climate.

TABLE 12.2

Population receiving natural high-fluoride water in Scandinavia, U.K., and U.S.A.

Country	Population receiving water containing appreciable natural fluoride	Fluoride level mg/litre
Denmark	95,000	1·0 or more[45]
Finland	144,000	1·0 or more[45]
Sweden	500,000	0·8 or more[45]
U.K.	482,000	1·0 or more[23]
U.S.A.	8,379,000	0·7 or more[40]

SUPPORT FOR FLUORIDATION

Fluoridation has been the subject of a number of official enquiries by government commissions and legal tribunals as well as by scientific bodies. After taking a great deal of evidence, Commissions in New Zealand,[29] South Africa[35] and Australia[1,8] have endorsed the safety and efficacy of the procedure, as did an official Canadian committee under the chairmanship of a Judge of the Ontario Court of Appeal[30] and also a judgement of the Supreme Court of the Republic of Ireland.[21]

In July 1969, in its twelfth plenary meeting, the World Health Organisation adopted the following resolution[44]

'To recommend Member States to examine the possibility of introducing and where practicable to introduce fluoridation of those community water supplies where the fluoride intake from water and other sources for the given population is below optimal levels, as a proven public health measure; and where fluoridation of community water supplies is not practicable to study other methods of using fluorides for the protection of dental health.'

This resolution has recently (1975) been re-affirmed[46] by the World Health Organisation at its 13th plenary meeting, which observed that fluoridation 'remains the most effective known means of preventing dental caries' and noted 'the further scientific proof of the safe use of fluoride in its various forms'.

OPPOSITION TO FLUORIDATION

There has been much opposition to fluoridation and the basis for this has been the subject of most of this report. In Sweden the Water Fluoridation Act of 1962 which enabled local authorities to request permission for fluoridation from the National Board of Health and Welfare was repealed in November 1971 by a vote of 137 to 126. The circumstances surrounding this decision have been described in detail by Burt and Petterson[6] but it may be noted that a previous attempt by opponents of fluoridation in 1968 to repeal this Act was heavily defeated by a vote of 254 to 54. Its repeal in 1971 followed much active campaigning by opponents and there is evidence that it was affected by party affiliations, 81 per cent of government members voting to retain the Act but only 18 per cent of the combined opposition.

In the U.S.A., a recent Federal Law[41] has declared that the addition of substances for preventive health purposes to water supplies is the responsibility of individual States and not of the federal government. This ruling has been quoted as evidence of a lack of support for fluoridation. In fact, no country has given so much support to fluoridation as the U.S.A. and almost half the population on public water supplies receive fluoridated water. This Act, in keeping with much federal legislation, declares that rulings on fluoridation should be the responsibility of individual States. This law, however, does require States to conform to national drinking water regulations which, in the case of fluoride, require that the level in water is below 1·4 to 2·4 mg/litre, depending on the annual average of the maximum daily air temperature.

It is sometimes incorrectly stated that Holland is opposed to fluoridation because the municipality of the Hague is mistaken for the Dutch Parliament. The Government of the

Netherlands is not opposed to fluoridation and, indeed, about 45 per cent of the population is served by fluoridated water. Similarly, statements that the procedure is illegal in the Federal Republic of West Germany are incorrect. In 1974 a law was introduced permitting individual States to request fluoridation. No example, apart from Sweden, could be found of a Government opposing fluoridation. In 1964 an Order was made in Denmark prohibiting the addition of fluoride to food or water. This was made on the advice of a Committee appointed by the Minister of the Interior which advocated fluoridation but was concerned to control fluoride supplementation preparatory to legislation enabling fluoridation of those water supplies with a low fluoride content. This legislation has not been introduced.

II. FLUORIDATION COMPARED WITH OTHER METHODS OF REDUCING CARIES

It is sometimes said that fluoridation is unnecessary since there are equally effective methods of achieving the same result. It is true that caries would decline if people stopped eating refined carbohydrates, but it cannot be expected that this would occur on a large scale.

Opposition to fluoridation and the lack of mains water supplies in certain areas have encouraged the study of other vehicles or methods of applying fluoride. Systemic methods include fluoride tablets and the use of milk, vitamin drops, salt or school water as vehicles while topical methods include fluoride-containing dentifrices, mouth rinses, and the application of solutions or gels to the teeth.

A. SYSTEMIC METHODS

(1) Fluoride Tablets

A number of studies and public health programmes have involved the daily administration to children of tablets containing 0.25 to 1 mg of fluoride and in at least one study, of 2 mg.[4,20] Although this method prescribes a specific dose, a major difficulty is that after an initial period of co-operation or even enthusiasm many children or their parents stop adhering to the regimen. This tendency can to some extent be overcome by school-based programmes, and several studies have demonstrated reduction in caries by these means.[20] Evidence from certain studies of a topical, post-eruptive effect, besides the systemic effect, suggests that these tablets should be sucked slowly to obtain the maximum benefit.

(2) Fluoridation of Milk

A reduction in caries following the addition of fluoride to milk has been reported from two small studies, one in Switzerland,[42,47] the other in Louisiana, U.S.A.[33] The concentrations of fluoride used in these studies were 1 mg/litre and 4 mg/litre respectively, reflecting the fact that the optimum level is not known. The method has the advantage of being selective since children under the age of 14 years would need to drink the fluoridated milk. However, there are several disadvantages. Although special 'dental milk' might be introduced into school milk schemes without too much difficulty, an important period of tooth formation takes place before school age. An additional category of milk would probably require modification of existing legislation to permit its sale under the name of 'milk'. Fluoridated milk would necessitate segregation and additional operations in the processing dairies besides special marketing arrangements which would increase the cost.

(3) Fluoride in Vitamin Drops

A controlled trial in Sweden[15] of fluoride in vitamin drops (A and D) has been shown to reduce the prevalence of caries. As a community measure this method is limited since there can be few areas where vitamin drops are taken on a large scale throughout the years of tooth development.

(4) Addition of Fluoride to Salt

Salt has been used as a vehicle for supplying fluoride in Switzerland and Hungary and has been shown to be effective at concentrations of 200–300 mg/kg in reducing caries.[27,38] Little is known about salt consumption at different ages and in different regions while it would be difficult to adjust to varying sub-optimal levels of natural fluoride in water supplies.[20]

(5) School Water Fluoridation

In the U.S.A. several studies[20] have demonstrated dental benefits in children from drinking water at school which was either fluoridated or contained appreciable levels of natural fluoride although they received negligible fluoride from the water at home. This procedure might, in some countries, have advantages in areas which lack a public water supply.

B. TOPICAL METHODS

(1) Fluoride-containing Dentifrices

Many studies have been carried out on the efficacy of dentifrices containing sodium or stannous fluoride or sodium monofluorosphosphate. These studies have usually measured the number of decayed or filled surfaces detected radiographically on teeth that have erupted during the study period. Using this measure, the reduction in caries reported by fluoride dentifrices has mainly been in the range of 15 to 30 per cent, being greater if their use is supervised. The question of fluoride ingestion by children who swallow some toothpaste has been considered by several workers.[11,13,14,18,19] It has been reported that approximately one quarter of the fluoride in a brushful of toothpaste was swallowed by children aged 4 to 7 years, amounting to 0·4 to 0·5 mg of fluoride.[11] Certain polishing agents in toothpaste have been shown to interfere with absorption of fluoride. A recent study[14] of 8 to 10-year-old children showed that on average approximately $120\mu g$ of fluoride was retained in the mouth after toothbrushing. The authors concluded that this small amount retained together with the limited absorption of fluoride from ingested toothpaste demonstrated the safety of fluoride dentifrices. There is evidence to indicate that in an area with 1 mg/litre of fluorides in the water, the use of these toothpastes results in a further reduction in caries, though this is slight.[32]

(2) Mouth Rinses

Some reduction in caries has been reported from the use of mouth rinses containing fluoride,[2] the effects being greater if used daily. This method is extensively used under supervision in Swedish schools.[37]

(3) Fluoride-containing Solutions and Gels

The application by dental personnel of a fluoride solution or gel to the teeth two or three times a year has produced reductions in caries, but has the disadvantage that it is a time-consuming procedure.[20]

COMPARISONS OF DIFFERENT METHODS OF GIVING FLUORIDE

It is extremely difficult to make a meaningful comparison between different methods of giving fluoride. Most fluoridation studies have covered several years (often over ten years) whereas studies of topical methods are usually carried out for only two or three years. Much of the reduction in caries reported by studies of topical fluoride have been detected radiographically and few have reported clinically significant differences. In practice, fluoride in the form of dentifrices or rinses is much less effective as a community measure in reducing caries than fluoridation. Fluoride tablets, drops, and fluoridated milk all require constant parental motivation and effort if their children are to receive an adequate supplement of fluoride. The application of fluoride solutions and gels is expensive in terms of dental manpower and it would seem that repeated applications are necessary if a worthwhile effect is to be maintained. Although the use of fluoride toothpaste is increasing, the important fact is that only a minority of the population use any type of toothpaste effectively and regularly.

The effect of utilising all approved dental health practices with the exception of fluoridation of the water supplies were studied over a ten year period in Askov, Minnesota.[22] The study began in 1948, and, besides a sustained programme of dental health education for both children and parents, involved topical application of a fluoride solution to erupted teeth, and supervised toothbrushing twice a day in the classroom. In addition, free toothbrushes and toothpaste were provided for home use, efforts were made to control excessive intake of sweets, and dental services were available on request. The authors concluded that the ten-year Askov dental health study cost more than fifty times as much as fluoridation, and was less than half as effective as fluoridation had proved to be in comparable communities.

III. ENVIRONMENTAL ASPECTS OF FLUORIDATION

The statement has sometimes been made that fluoridation would constitute pollution of the environment because of the large amounts of fluoride that would be discharged into rivers and into the sea. It can be estimated that fluoridation of all public water supplies in England, Wales and Scotland would result in the discharge of approximately 6,000 tonnes of fluoride annually.

If this quantity were to be distributed evenly in rivers the average fluoride concentration would be approximately 0.14 mg/litre in England and Wales and 0.03 mg/litre in Scotland.[5,36,39] However, such a uniform distribution could not occur: those rivers flowing through major population centres would contain higher levels while many smaller rivers would have lower levels. Some recycling of fluoride would occur from river abstraction sources situated downstream from discharge points and this would reduce the amount of fluoride that would have to be added to raise the level to 1 mg/litre in water supplies. The highest fluoride level in any river as a result of fluoridation would be appreciably less than 1.0 mg/litre whereas there are rivers such as the River Brazos and Canadian River in the U.S.A. that contain levels of over 6 mg/litre. A review of the effects of fluoride on plant and animal life concluded that in the case of aquatic life of all forms, a concentration of 1.5 mg/litre has no unfavourable effects.[26]

Fluoridation would have a negligible effect on the fluoride concentration in the sea, which is 0.8 to 1.4 mg/litre, not only because of the enormous dilution that would occur but because most of the fluoride added to water supplies would be derived from sources that would otherwise have been discharged to the sea as waste.

The toxicity of certain plants in Australia, Africa and South America long known to be dangerous to animals has been shown to be due to their ability to synthesise fluoroacetate.[7,31] This fact has prompted the suggestion that fluoridation might in some way result in the formation of toxic amounts of such substances in cereals, fruits and vegetables—although this would appear to overlook the fact that fluoride is abundant in soil naturally (a typical soil contains 200 mg/kg).[45] It has been shown that some forage plants treated with inorganic fluoride convert very small amounts of this to fluoroacetate and fluorocitrate,[25] but the amounts are very small and well outside the toxic range.[7] Such small amounts of fluorocitrate are already common constituents of food.[31]

IV. ECONOMIC CONSIDERATIONS

The cost of building and running a fluoridation scheme depends not only on the population covered but also on the complexity of the water supply arrangements that will determine the number of fluoridation plants required. Thus, the annual cost in Birmingham with a large population and a straightforward supply system was estimated recently at 1·5p per head, but in certain other areas the costs are greater. The estimated costs have recently been published for schemes in several countries by Davies,[9] the annual cost per head, together with an allowance for capital outlay for equipment, ranging from 1·25 to 2·56p in the U.K. and 2 to 11 cents in the U.S.A. It is impossible to assess the financial value of preventing, say, the pain of caries. Nevertheless, several attempts have been made to measure the economic consequences of fluoridation in terms of either reductions in the staff required to provide dental service or in savings in the cost of dental treatment. It was recently estimated that if a saving of one decayed, missing or filled tooth is assumed to be equivalent to the saving of two 2-surface amalgam fillings costing £2·70 at current fees, the savings in the cost of fillings at Watford which was fluoridated in 1955, ranged from £1·62 per child at the age of 3 years to £4·32 per child at the age of 7.[9] All the indications are that even apart from the obvious social benefits, fluoridation would produce appreciable savings in the overall costs of the dental service.

It is recognised that a large proportion of fluoridated water is not drunk by children under the age of 14 years, and in this sense fluoridation has been said to be uneconomic. However, in view of the low cost of fluoride and the inadequacy of alternative methods of providing this on a community basis, fluoridation is economic.

REFERENCES

1. Australia (Victoria) (1954). *Report of the Special Committee appointed by the Commissioner of Public Health.*
2. Birkeland, J. M., Jorkend, L. and Von Der Fehr, F. R. (1973). *Community Dent. and Oral Epidemiol.*, 1, 17.
3. Black, A. P. (1962). *J. Am. dent. Ass.*, 65, 588.
4. *British Medical Journal* (1976). Editorial, 1, 535.
5. British Water Supply Year Book (1974). *Official Year Book of the British Waterworks Association.* Wheatland Journals Ltd.
6. Burt, B. A. and Petterson, E. O. (1972). *Brit. dent. J.*, 133, 57.
7. Ciba Foundation (1972). *Carbon-Fluorine Compounds.* N. Holland: Excerpta Medica.
8. Crisp, M. P. (1968). *Report of the Royal Commissioner into the fluoridation of public water supplies.* Hobart, Tasmania: Govt. Printer.
9. Davies, G. N. (1974). *Cost and Benefit of Fluoride in the Prevention of Caries.* W.H.O. Geneva.
10. Dunning, J. M. (1965). *New Eng. J. Med.*, 272, 30 and 84.
11. Ericsson, Y. and Forsman, B. (1969). *Caries Research*, 3, 290.
12. Forrest, J. R. (1967). *Brit. dent. J.*, 123, 269.
13. Forsman, B. and Ericsson, Y. (1973). *Community dent. Oral Epidemiol.*, 1, 115.
14. Glass, R. L., Peterson, J. K., Zuckerberg, D. A. and Naylor, M. N. (1975). *Brit. dent. J.*, 138, 423.
15. Hamberg, L. (1971). *Lancet*, i, 441.
16. Hansard. 25 June (1975). Col. 149.
17. Hansard. 28 Jan. (1975). Col. 149–150.
18. Hargreaves, J. A., Ingram, G. S. and Wagg, B. J. (1970). *Caries Research*, 4, 256.
19. Hargreaves, J. A., Ingram, G. S. and Wagg, B. J. (1972). *Caries Research*, 6, 237.
20. Horowitz, H. S. (1973). *Community dent. Oral Epidemiol.*, 1, 104.
21. Ireland, Republic of (1963). *Fluoridation: Judgement delivered . . . in the High Court, Dublin.* Dublin, High Court.

22. Jordan, W. A., Snyder, J. R., Peterson, J. K., Johnson, R. A. (1959). *North-West Dentistry*, **38**(6), 445.

23. Kinlen, L. J. (1975). *Brit. dent. J.*, **138**, 221.

24. Kunzel, W. (1974). *Caries res. Suppl.*, **8**, 28.

25. Lovelace, C. J., Miller, G. W. and Welkie, B. W. (1968). *Atmosph. Environ*, **2**, 187.

26. McKee, J. E. and Wolf, H. W. (1963—reprinted 1971). *Water Quality Criteria.* Californian State Water Resources Control Board, Pub. No. 3-A.

27. Marthaler, T. M. (1967). *Int. dent. J.*, **17**, 606.

28. Ministry of Health (1953). *The fluoridation of domestic water supplies in N. America as a means of controlling dental caries: Report of the United Kingdom Mission.* London: H.M.S.O.

29. New Zealand Government (1957). *Report of the Commission of Inquiry on the Fluoridation of Public Water Supplies.* Report No. H47. Wellington, N.Z. Govt. Printer.

30. Ontario Government (1961). *Report of the Committee appointed to enquire into, and report upon, the fluoridation of municipal water supplies.* Toronto.

31. Peters, R. In Ciba Foundation (1972). *Carbon-Fluorine Compounds.* N. Holland: Excerpta Medica. Pages 1 and 64.

32. Peterson, J. and Williamson, L. (1975). *J. dent. Res.*, **54**, l. 85.

33. Rusoff, L. L., Konikoff, B. S., Frye, J. B., Johnston, J. E. and Frye, W. W. (1962). *Amer. J. clin. Nutr.*, **11**, 94.

34. Scobie, R. B. (1971). *Ala. J. med. Sci.*, **8**, 439.

35. South Africa, Republic of (1966). *Report of the Commission of Inquiry into Fluoridation.* Pretoria. Govt. Printer.

36. *The Surface Water Yearbook of Great Britain 1966–70* (1974). Water Resources Board and the Scottish Development Department. London: H.M.S.O.

37. Torell, P. and Ericsson, Y. (1965). *Acta Odontologica Scand.*, **23**, 287.

38. Toth, K. (1973). *Caries Res.*, **7**, 269.

39. Twort, A. C., Hoather, R. C. and Law, F. M. (1974). *Water Supply.* (2nd edn.). London: Edward Arnold.

40. U.S. Department of Health, Education and Welfare (1970). *Fluoridation Census 1969.* Washington.

41. U.S. Government Printing Office, Washington D.C. December 16 (1974). *Safe Drinking Water Act.* Public Law 93-523.

42. Wirz, R. (1964). *Schweiz. Monats. fur Zahnheilkunde*, **74**, 767.

43. World Health Organisation Chronicle (1969), **23**, 505.

44. World Health Organisation (1969). *Twelfth Plenary Meeting.* Resolution W.H.A. 22.30.

45. World Health Organisation (1970). *Fluoride and Human Health.* Geneva.

46. World Health Organisation. 28th World Health Assembly. *Thirteenth Plenary Meeting.* W.H.O. 28.64.

47. Zeigler, E. (1964). *Helvetica Paediatrica Acta*, **19**, 343.

13 Conclusions

Water containing fluoride at a level of 1 mg/litre has been drunk for generations by millions of people throughout their lives. Since fluoridation was introduced, millions more have been drinking water with this level for many years. In both situations, medical and radiographic surveys have been carried out, sometimes also covering areas with up to 8 mg/litre of fluoride. Area mortality rates have also been analysed in terms of their water-fluoride levels. There is now an enormous body of information bearing on the subject of fluoride and health which amply justifies the following conclusions—

1. Fluoride in water added or naturally present at a level of approximately 1 mg/litre over the years of tooth formation substantially reduces dental caries throughout life.
2. There is no evidence that the consumption of water containing approximately 1 mg/litre of fluoride in a temperate climate is associated with any harmful effect, irrespective of the hardness of the water.
3. In comparison with fluoridation, systemic fluoride supplements such as tablets, drops and fluoridised salt have not been shown to be as effective on a community basis.
4. There is no evidence that fluoridation has any harmful environmental effect.

14 Recommendation of the College

The College recommends fluoridation of water supplies in the United Kingdom where the fluoride level is appreciably below 1 mg per litre.